HOW TO TALK ABOUT CLIMATE CHANGE IN A WAY THAT MAKES A DIFFERENCE

Praise for *How to Talk About Climate Change in a Way That Makes a Difference*

'Explains whether and how we will choose to solve the climate problem. Immensely important analysis in a great read.'
—Professor Ross Garnaut

'Rebecca Huntley has given us a great gift: an essential guide to understanding ourselves and each other as we face the climate crisis. Let's take down the walls that divide us. Collectively, with compassion and courage, we can make real change happen.'
—Kylie Kwong

'A book on how to talk about climate change is merely another reading option for the committed. A book about being vulnerable, facing fear, despair as well as guilt that moves to hope, love and naming the deeply felt things is compulsory reading. This is one such book. Rebecca Huntley, esteemed for her objective surveys that tell us what ordinary Australians think, jumps a fence. The detached observer practises her craft by becoming the subject of her craft. Her vulnerability is not indulgent sentimentality but works to chart the risks that climate change brings to the people and things she loves most in the world. In so doing she powerfully reminds me of what is most at stake. Indeed, this book does us all a great service.'
—Tim Costello, Executive Director of Micah Australia

HOW TO TALK ABOUT CLIMATE CHANGE IN A WAY THAT MAKES A DIFFERENCE

REBECCA HUNTLEY

murdoch books
Sydney | London

Published in 2020 by Murdoch Books,
an imprint of Allen & Unwin

Copyright © Rebecca Huntley

All rights reserved. No part of this book may be reproduced or transmitted in any form or by any means, electronic or mechanical, including photocopying, recording or by any information storage and retrieval system, without prior permission in writing from the publisher. The Australian *Copyright Act 1968* (the Act) allows a maximum of one chapter or 10 per cent of this book, whichever is the greater, to be photocopied by any educational institution for its educational purposes provided that the educational institution (or body that administers it) has given a remuneration notice to the Copyright Agency (Australia) under the Act.

Murdoch Books Australia
83 Alexander Street, Crows Nest NSW 2065
Phone: +61 (0)2 8425 0100
murdochbooks.com.au
info@murdochbooks.com.au

Murdoch Books UK
Ormond House, 26–27 Boswell Street, London
Phone: +44 (0) 20 8785 5995
murdochbooks.co.uk
info@murdochbooks.co.uk

 A catalogue record for this book is available from the National Library of Australia

A catalogue record for this book is available from the British Library
ISBN 978 1 76052 536 1 Australia
ISBN 978 1 91163 276 4 UK

Cover design by Trisha Garner
Text design by Susanne Geppert
Typeset by Midland Typesetters
Printed and bound in Australia by Griffin Press

Grateful acknowledgment is made to the following for permission to reprint quoted material:

Don't Even Think About It by George Marshall, copyright © George Marshall, 2014, Bloomsbury Publishing Inc.

Losing Earth by Nathaniel Rich, copyright © Nathaniel Rich 2019. Reproduced with permission of the Licensor through PLSclear.

Image on page 59 provided by the Australian War Memorial (ARTV00433). Original poster by Savile Lumley, UK Parliamentary Recruiting Committee, 1915.

Every reasonable effort has been made to trace the owners of copyright materials in this book, but in some instances this has proven impossible. The author(s) and publisher will be glad to receive information leading to more complete acknowledgements in subsequent printings of the book and in the meantime extend their apologies for any omissions.

10 9 8 7 6 5 4 3 2

 The paper in this book is FSC® certified. FSC® promotes environmentally responsible, socially beneficial and economically viable management of the world's forests.

To all the kids who went on strike for the climate and to my own kids, Sofia, Sadie and Stella

CONTENTS

INTRODUCTION
A CHANGE OF HEART
Or how I became emotional about climate change
1

1
THE PROBLEM WITH REASON
Or why we need to stop arguing about science
13

2
START BEING EMOTIONAL
Or the importance of feelings over facts
29

3
GREEN GIRLS
Or what we can learn from teens on talking about climate
45

4
GUILT
Or my plastic coffee cup killed the green sea turtle
67

5
FEAR
Or do wildfires change minds and votes?
91

6
ANGER
Or how to turn anger into activism
115

7
DENIAL
Or the need to be innocent
131

8
DESPAIR
Or the support group at the end of the world
151

9
HOPE
Or how to get out of bed in the morning
177

10
LOSS
Or bury me in a carbon sink
195

11
LOVE
Or do it for the birds
217

CONCLUSION
TALK ABOUT CLIMATE CHANGE
It's the right time
231

Acknowledgements **249**
Notes **251**
Suggested reading and resources **271**
Index **277**

INTRODUCTION

A CHANGE OF HEART

Or how I became emotional about climate change

On 1 December 2018 I woke up early as usual, before my children stirred from their beds, made myself a cup of coffee and turned on the TV. I switched straight to my favourite early morning news and current affairs show and I saw them. Hundreds of Australian teenagers skipping school the day before and protesting in the streets about climate change.

They were holding handmade signs with slogans that ranged from the serious and angry to the humorous and profane. 'There are no jobs on a dead planet.' 'Act now or swim later.' 'You're burning our future.' In Australia in the weeks leading up to the strike, the prime minister admonished the children for protesting rather than staying in school. My favourite sign of the protest was a direct response to his rebuke: 'Why should we go to school if you won't listen to the educated?'

As I sat sipping my coffee, I thought to myself, 'Good on those kids telling the powers that be, the older generation, that they need to do more about climate change.' And then it hit me.

At almost 50 years of age, I *am* part of that older generation, part of the generation in power, with the platform and a voice these young people don't have. It was, at that moment, as if those teenagers, their signs both funny and grave, were speaking to me. We need you to listen. We need you to act.

And just then something shifted inside me. It was a sensation that's hard to describe, and yet I can recall it now with clarity. It actually felt *physical*, as if vital organs had moved inside my body, the way they did after I gave birth to my children. Those truant teens were telling me to do something. And I said back to the TV screen, out loud (although only our dog would have heard it), 'Okay.'

I know I was not alone in my generation in reacting the way I did; so many adults have been 'schooled' by the strike, inspired to think about the world differently. From that point on, my life changed. In my social research career, I started choosing to work with more and more environmental organisations. In my broadcasting and writing, I gravitated towards environmental issues. I applied for and was accepted to do the three-day Climate Reality training course in Brisbane, with 800 other people from across the Asia-Pacific region. I started reading every book I could find on climate change. I switched all our home energy to renewables. I changed the investment mix of my retirement fund away from fossil fuel companies. I sourced a butcher in the area to provide my family with sustainable meat. I started researching hybrid and electric cars. Sitting in a restaurant with some friends a few months afterwards, I ordered a vegetarian meal. 'I'm cutting down on meat for environmental reasons,' I explained. 'Something real has happened to you, hasn't it?' said one of my dinner companions. It had.

Cynics might say I am describing a kind of road to Damascus moment, placing me among the ranks of zealots on climate who approach the issue with religious devotion. And yet almost all environmental activists have gone through this transition from general concern about the environment to passionate connection to its protection, a kind of calling that will direct their lives in all kinds of ways. The only difference seems to be that the catalyst for this transformation varies depending on the person.

Looking back on this moment, what's obvious to me—and fascinating—is that watching those young protestors on the streets didn't mean I suddenly believed the scientific consensus on climate change more than I had done the day before the climate strike. For as long as I had known about climate change, I had believed it was happening and that it posed a serious future threat.

Concern about climate change influenced my voting choices. I wanted governments to act. Certainly, I had always bought environmentally friendly products, recycled and composted. I offset my flights. I even donated to environmental causes. I saw both Al Gore documentaries in the cinema. I wouldn't have described myself as an environmentalist, but I did see myself as environmentally conscious. It's clear, though, that environmental concern was not at the centre of my worldview. I had registered it rationally but not emotionally.

This emotional change intrigued me. I consider myself a highly rational person. I'm a trained lawyer and social researcher. I base my judgements on demonstrable evidence that will stand up to scrutiny from lawyers, good journalists, academics and Senate committees. But this transformative moment—the moment I tipped from being concerned about climate change to genuinely alarmed about the threat—didn't happen because I read a report from the Intergovernmental Panel on Climate Change (IPCC) or sat through a presentation from a climate scientist about carbon dioxide levels. I reacted to a crowd of children holding up signs in the streets, girls who were only a few years older than my eldest daughter. Suddenly, it was personal.

Perhaps I was always headed in this direction; the slow drip of information about climate change over the years and my politically progressive tendencies left me open to the transformation I've described to you. And there were flashes of feeling that foreshadowed this change of heart in the previous years. Interviewing Australian farmers about climate change for a radio documentary, I couldn't help but be affected by their stories of the hardships of farming in a climate age—third- and fourth-generation farmers anxious that they would never be able to pass their land on to their children or grandchildren. Snorkelling on the Great Barrier Reef with my family, I glanced across the water through a school of blue and grey fish as my then six-year-old surveyed the scene with wild delight. And I wondered, will she do the same thing with her daughter?

Or will there be nothing for them to see but bleached coral, a ruin and not a living thing? (They call this reef grief.)

But despite these isolated moments over the last decade or two, it took a kids' street protest to bring about a more substantial change. I am now on an endless emotional roller-coaster, shifting from fear to anger, sadness, hope, acceptance and even now and then a form of denial (perhaps it won't be as bad as they say . . .). Some nights I lie in bed worrying about the future of my children in a climate-altered world. What can I do to prepare them? And what will I say if they ever ask me, 'Why didn't you do more?'

While I'm not in the majority in my own country or most countries around the world in terms of how I feel about climate change, I'm not the only one shedding tears over the children's climate strike. A few weeks after my 'Okay' moment in front of the TV, I read an opinion piece online that showed my experience was in no way unique. It was written by the CEO of a regional not-for-profit organisation, Rob Law, who has studied and worked on climate change as a job for over fifteen years. He observed that his approach to climate change was becoming less 'intellectual and rational' and much more emotional and personal:

> There have been several instances lately where the impacts of climate change have hit me hard and unexpectedly. Perhaps this is because I am now a father or because scientific projections I learned about 15 years ago are now unfolding quicker than imagined.

In March I went to my local train station to watch 500 schoolkids gather to commute to Melbourne for the big school strike. I was surprised that I found myself moved to tears and overcome with emotion, and that I wasn't alone among the other adults there . . .

What I find curious is that . . . these instances were essentially positive, inspiring moments. Yet they seemed to bring forth sadness or internalised grief that had been buried out of sight.[1]

Like everything in life, when it comes to your feelings about climate change, it helps to know you're not alone.

You won't find a lot of natural science in this book. Not a lot of data about CO_2 levels and temperatures rises, shrinking ice sheets and acidification of the oceans. This is a book about humans, drawing on the social sciences—psychology, sociology and even evolutionary psychology.

It's supposed to be a kind of self-help guide for the climate age (minus the survival tips or advice about managing eco-anxiety). A way to help you better understand how the people around you are reacting to climate change. How we, as humans, respond to the information (and misinformation) we get from the media, from politics, from scientists and from society at large. And, increasingly, how we respond to our real-life experience of climate change, including temperature rises and weather events.

In this book, I explore and touch on the whole spectrum of emotions—everything from anger and fear to love and loss. While climate change can evoke many of these emotions at the same time, I deal with them one by one, starting with guilt and ending with love.

I hope that by reading this book you'll get a better understanding of your own reactions to the issue—and how to manage the roller-coaster I've described above. The aim of this book is unashamedly an activist one. I want you to read it and become better equipped to talk about the climate with the people around you and encourage them to act. And be able to sustain that over time, as news and events associated with climate change get more confusing, emotionally disturbing and politically divisive.

You might ask, 'Why does it matter what we, as individuals, feel about climate change? Or even as groups of people? Isn't it more about governments and corporations, those with the most power, acting to do something about it? Aren't our feelings beside the point?' While I agree that we need governments and powerful economic interests to act, there is no doubt we need multiple fronts of pressure on governments and corporations, especially those that are resisting action.

Understanding the social and psychological factors that underpin how we respond to climate change is the first essential step to understanding how we can better persuade people to act. That action includes voting out or protesting against governments and corporations that refuse to take climate change seriously.

While I said this isn't a book with a lot of facts and figures about climate science, I do need to devote a few sentences to the consensus on climate change—and by consensus I mean that 99 per cent of experts in the area agree this is actually happening. The rise in the global temperature caused by pollution since the Industrial Revolution is destabilising our weather. This heat is melting the ice shields in Greenland and Antarctica, causing sea levels to rise and threatening low-lying Pacific countries and parts of Australia, as well as coastal communities around the world. Extreme weather events like storms, cyclones and wildfires as well as droughts are more common now. We are also in the middle of the world's sixth mass extinction, the worst since the time of the dinosaurs, driven by the changing climate as well as deforestation and land clearing.

Climate change is affecting every one of the earth's systems. Even the earth's crust is being transformed. While making predictions about the future is always tricky, the consensus is we have just over a decade to cut pollution or CO_2 and limit the global temperature rise to 2 degrees Celsius or less if we are to prevent the collapse of the natural world and the societies that rely on it. Many scientists and experts in the field of national security say this process is already underway. The extreme weather events and sea level rises we are observing around the world are happening at a 1 degree increase, and we are tracking towards overshooting 2 degrees and moving towards 3 degrees. The impact of climate change is already being felt in our food chains, contributing to refugee crises in some countries, and creating a global security nightmare for domestic governments and international bodies.

I hope I haven't lost you! This is usually the moment in a conversation about climate change where people react very differently. If, like me, you're already convinced and highly concerned about the climate, you nod your head at this information and reach for a vodka and tonic regardless of the hour. If you're a climate sceptic or denier, you gear up to fight the veracity of these statements; you might also feel like you need to shoot the messenger—'Huntley, you're an inner-city, Al Gore–loving leftie. You would say that!'

But if you're not a convert or a denier, your reactions could vary enormously. You could put this book down and never pick it up again because it's all too depressing and overwhelming. Please don't do that, stay with me! Or you could feel confused and doubtful—'How can they say this is happening? How do they really know?' Or it could all leave you cold—'I've heard this before. They said the Y2K bug would be the end of the world.' Or you could think, 'I believe all this and I want to do something about it. But what can *I* do?'

If you're still reading, I'm glad and grateful. Trust me, this book is supposed to encourage you to be hopeful. Not the kind of hope that's rooted in a Pollyanna-like belief that everything will be okay as long as we separate our recycling and ride a bicycle to work. This kind of hope has quicksand as its foundations.

The hope I want to encourage is based on the belief that we have the technology to address the climate challenge and we can come up with a human response to this human-caused threat. That we can find more sophisticated ways to communicate the threat to all kinds of people in all kinds of societies, and respond in a way that will salvage the future.

This is a hope based on action. Greta Thunberg, the Swedish teenager whose personal strike for the climate inspired the global student strikes, put it this way in her TED talk: 'Once we act, hope is everywhere.'[2] And sometimes that hope-generating action can just be meaningful words exchanged with another person.

One of the greatest sources of optimism for me is listening to the stories of people who are doing great things in their local community to convince others that climate change is real and they need to act. In his book *What We Think About When We Try Not to Think About Global Warming: Toward a new psychology of climate action*, Norwegian psychologist and economist Per Espen Stoknes writes about the critical role of stories in making climate change seem personally relevant and less overwhelming.

He makes the point that as human beings we rely heavily on storytelling to help create meaning in our lives, and forge personal and collective identities. He also says that the dominant story told about climate is one of 'the apocalypse of climate hell',[3] almost biblical in dimensions, one of famine and destruction, burning rivers where we live in *Mad Max*–like societies killing each other for a litre of fresh water.

Stoknes reminds us that this story, by generating fear, guilt, anger, despair and helplessness, allows climate denialists to ridicule environmentalists as somewhat deranged, like people standing on a street corner wearing a sandwich board saying 'The end of the world is nigh'. For anyone who isn't already seriously considering such a future, it's an approach that can turn people off and shut down conversation. But Stoknes rightly points out that this CO_2-generated hellscape is just one vision of the future we can put forward.

> I don't think there is just *one* right type of climate story to tell to get people to understand the urgency of the issue and move them to action. Rather, a plurality of stories is needed, each creating meaning and engagement for different groups of people.[4]

He argues for more stories about 'what's going well, of conviction and endurance, as well as stories that describe and help us imagine a renewal of society, wildlife and ecosystems'. Stories of 'people who care and act on the basis of vision, determination and joy'.[5]

This book shares many of these stories from people who aren't the most famous voices in the climate change fight, but that's been a deliberate choice of mine—and David Attenborough didn't return my calls. It's these different stories of different kinds of people from different cultures, stories of action, that are more likely to move people. Stories from people who look and sound like we do.

It was the actions of children like my own that brought about my change of heart and began a new era of climate-related work for me. All the chapters in this book open or end with interviews with people who are active in the climate change cause, and have a story to tell about their involvement and advice to give about how to talk about climate change effectively.

In her TED talk, Greta Thunberg asks us to wake up and change.[6] That Saturday after the climate strike, I literally woke up and changed. My heart changed years after my head understood how important an issue climate change is to my future and the future of my kids.

What will bring about a change of heart in others who aren't like me, who don't have the same ideas about the world I do, is now the central challenge in the fight to save the planet. It's a problem not of science and technology but of how we communicate and how we encourage action. It will be different for different people in different cultures, but if it's successful the result will be the same. A shared world saved, a future for all children, not just mine.

Author's note
As this book went to press, many parts of the world were experiencing unprecedented levels of disruption with the COVID-19 pandemic. It's impossible to predict how things will unfold over the coming months, but we are clearly living through another challenge requiring previously unthinkable shifts in behaviour and levels of international cooperation. Perhaps this will offer a path forward for the climate challenge as well. The larger issues in this book remain unchanged, and I have inserted connections to the pandemic where relevant.
—Rebecca Huntley, April 2020

CHAPTER 1

THE PROBLEM WITH REASON
Or why we need to stop arguing about science

It took me much longer than it should have to realise that educating people about climate change science was not enough. Due perhaps to my personality type (highly rational, don't talk to me about horoscopes, please) and my background (the well-educated daughter of a high school teacher and an academic), I have grown up accepting, albeit unconsciously, the idea that facts persuade and emotions detract from a good argument.

Then again, I'm a social scientist. I study people. I deal mostly in feelings not facts. A joke I like to tell about myself during speeches is that I'm an expert in the opinions of people who don't know what they're talking about. Over the fifteen years I've been a social researcher, I've watched with concern the increasing effects of climate change in my own country, and also watched as significant chunks of the electorate voted for political parties with terrible climate change policies.

There is clearly a disconnect between what people say they are worried about and want action on and who, when given the chance, they pick to lead their country. You can rationalise that all you like (the surveys are wrong, people care but they don't trust politicians), but the reality is that people can say one thing and mean it in one context but do something that contradicts that concern in another context. We can be consistently inconsistent. My training and persona mean I'm attracted to reason and facts, but my stock in trade is emotion and the 'irrational' arguments people use to justify what they do.

It's clear too that I've grown up tacitly accepting that the information-deficit model of communication, particularly when it comes to science, is the correct way to proceed. By the 'information-deficit model', I mean a one-way communications style, where 'scientists and experts' relate information directly to a 'clueless public'. This information can sometimes be delivered in complex language, using jargon and presuming a certain level of understanding of specific concepts. The experts tell the proles the facts, the proles trust the experts, accept the information and act accordingly.

Clearly there are many, *many* flaws in this approach. It assumes a strong level of understanding of scientific concepts in the community at large, which is a shaky assumption. The Programme for International Student Assessment (PISA) world rankings for maths and science education show that many countries, including affluent and so-called developed countries, have a long way to go. This approach also assumes a general trust of experts and expertise in the community. Of course, in countries like the United States, Britain and even my home country Australia, 'experts' have been attacked by some politicians and media figures as being 'out of touch' with the community. Their cred is under a cloud. The widespread collapse in trust in our leaders and our institutions has impacted the academy and other institutions of expertise as well.

But the biggest flaw in this approach to science communication of course is that it ignores the fact that we are dealing with human beings.

Yes, human beings. Pulitzer Prize winner and Harvard professor Edward O. Wilson describes us in these terms in his compelling book *Half-Earth*:

> Thinking with a gabble of reason, emotion and religion . . .
> Magnificent in imaginative power and exploratory drive,
> yet yearning to be more master than steward of a declining
> planet. Born with the capacity to survive and evolve
> forever. Arrogant, reckless, lethally predisposed to favor
> self, tribe and short-term futures.[1]

Reason and emotion. Imaginative and reckless. Constantly evolving and selfish. Loyal to tribe above almost all else. Any approach to communicating about climate change that doesn't take these human characteristics into account will not be very effective.

Per Espen Stoknes argues that the stubborn influence of the information-deficit model of science communication has undermined the progress of the climate movement. He contends that in believing more facts will convince more people of the need to act, we ignore messy social reality. Instead, we should be guided by what neuroscience and psychology can teach us about how our brains actually work and respond to climate change messages.

> Just presenting such facts and figures about global warming over and over again has so far *not sufficiently* convinced the general public, journalists, or policy makers about the scale of the problem to create the sense of urgency needed for required actions . . . Selling the science is a mightily difficult task, since the issue of climate change is conveyed as remote, abstract, vague . . . You might even call it ghostly or diabolical since it seems to defy neatly all the evolutionary and cognitive hooks our brains use to generate a sense of urgency.[2]

More on the issue of fear and urgency later in this book, but suffice to say that in the research I do in Australia on how people respond to climate change messages, I've observed first-hand the limits of the 'facts and figures' approach to climate communication. In the last couple of years in particular I've had people in focus groups who, when presented with the basic facts of climate change, including that we're on track for catastrophic warming, respond with '[*Yawn*] We've heard all this before'. The facts and figures of climate change have become like a familiar script that can leave us cold or, even worse, bored.

'The greatest science communication failure in history.'[3] That's how Stoknes describes the too slow rise in public concern and action since climate change was 'discovered' in the late 1970s.

Public concern has not mirrored the scientific case that the climate is changing. In fact, in some parts of the community, the outcome has been: the more facts, the *less* concern. To top it off, instead of being heeded, climate change messengers have encountered both 'vicious counterattacks' and 'impenetrable walls of psychological backlash or indifference'.[4]

Journalist Nathaniel Rich in his book *Losing Earth* puts it another way. Over the last three decades, more and more of the scientific predictions about climate change have panned out and more scientific consensus has built up to an almost unanimous view that climate change is being caused by human activity. And while there have certainly been shifts during this time in public attitudes to the issue, community resistance to the facts or outright denial have not melted away in the face of expert opinion.

The science is solid and well circulated, and yet the arguments and division in the community remain. As Rich writes, 'there has been no fundamental change in climate physics since 1979, only refinement' and yet 'nearly every conversation that we have [now] about climate change was being held in 1979'.[5] He concludes soberly, 'it no longer seems rational to assume that humanity, encountering an existential threat, will behave rationally'.[6]

To make matters worse, the nature of scientific inquiry itself is also a barrier to lay people accepting the science. The scientific method involves a particular tried and true approach, namely the observation of phenomena, the formulation of a hypothesis concerning those phenomena, experimentation to demonstrate the truth or falseness of the hypothesis, and a conclusion that validates or modifies the hypothesis. Criticism is the backbone of the scientific method, and something can only be determined as fact if it has been put through an intensive process of peer review and criticism. Even when scientists arrive at a conclusion, there are often so many caveats attached to it that it can seem to a non-scientist as less than solid and reliable.

Combined with the ability of powerful elements in the global media to make out as if climate science is still 'in doubt' or 'up for debate', the very nature of the scientific method means that people without scientific training feel even more confused by the 'rational' arguments presented to them. (I will return to the idea of doubt and uncertainty in the chapter on denial.) Lay people can see 'uncertainty' in scientific arguments as much more problematic than scientists do themselves.

It's not just that scientists emphasise uncertainty as part of their thinking. British science communicator George Marshall argues that given the nature of their training, scientists find it hard to present their research findings in ways that connect to people's lives. 'They often excise the very images, stories, and metaphors that might engage our emotional brain and galvanize us into action,' he writes.[7] This includes personal storytelling, which has a powerful ability to engage the emotional brain and convince us that the messenger really cares about what they're telling us.

The whole culture of science works against the inclusion of the personal in communicating research. In my work with academics, especially scientists, I often detect a degree of unease when it comes to talking about what matters to them as human beings rather than just as experts. To include personal stories and motivations feels like 'dirty talk'.

I have always pondered the irony that climate change requires politicians to talk about science and scientists to grapple with politics. I feel sorry for climate scientists, to tell you the truth, and perhaps we're being unfair when we criticise them for believing that facts will win the argument and a one-way approach to science communication is the right way to go. I suspect many of them got into their jobs so they could hang around the laboratory with other data nerds. None of them, I imagine, yearned for a career going head to head with politicians and pundits on live TV panel shows, where their well-researched ideas are countered by outright lies and distortion for the purposes of infotainment.

While famous and skilled science communicators like David Attenborough, David Suzuki, Bill Nye, Brian Cox and Neil deGrasse Tyson are often singled out by the public as some of the most reputable messengers on climate change, there's a limit to their capacity to shift public opinion. Knowledgeable dudes talking on TV about space or wetlands or the breeding habits of sea lions are all very interesting, but the problem of connection still remains. How does this relate to my life?

Climate scientists in particular realise that when it comes to the climate change cause, the bulk of their work has been done. All they are doing is updating the data on a theory already proven countless times to be true.

They know the important steps now are social, political, cultural and economic.

Let me a share a story to illustrate this point. As I mentioned in the introduction, I was lucky enough to attend the Climate Reality conference in Brisbane in 2019. Climate Reality was started by former US vice-president Al Gore as a global network that educates people from all walks of life to talk about climate change in their communities. As part of the three-day training, a panel of climate scientists was talking about the latest science. Al Gore himself ran the panel and posed the questions.

One of the panellists was an expert on the Great Barrier Reef, the world's largest coral reef system, a World Heritage site under serious threat of destruction through temperature rises and acidification causing mass coral bleaching events. The former vice-president asked this esteemed research scientist what he would do if he was suddenly gifted 1 million dollars. What kind of research project would he want to undertake on the reef system that had been his passionate interest for so many years?

The climate scientist replied that he wouldn't put the money towards any further scientific research but into events like the one we were holding in Brisbane, where 800 concerned people from around the country and the Pacific were assembled to learn more about climate change and how to talk about it with other people. An extraordinary admission for a research scientist (especially in a country where research grants are hard to obtain), and a reflection that the main challenge now is not to unearth more facts about climate change but to find a way to make all of us messy, clever and contradictory humans act to save our planet.

It's not as if the inadequacy of the information-deficit approach to communication is news to environmental groups. The many environmental organisations I work with have for some time been thinking more and more about how to appeal to citizens and consumers based less on rational and expert arguments and more on values and worldview. This is the case around the world.

That being said, it's amazing how often climate communicators forget this important principle. A couple of weeks after I decided to write this book, I went to see a climate scientist speak at a regional arts festival. The festival venue was packed with fans buying up books on the environment, not a plastic bag or plastic straw in sight. The speaker had the obligatory PowerPoint deck of a researcher with a PhD, with what must have been over 100 slides. I figured he spent at least two-thirds of his allotted hour and a half going through slides with complex graphs and statistics about CO_2 levels and sea level rises, the aim of which was to convince—or further convince?—the crowd that climate change was happening. And yet this was the epitome of preaching to the converted.

If *these* people from *this* community were giving up an afternoon to come to *this* talk, then they were already on board. What they really needed was not facts but solace. Watching them, listening to their comments before and after, and during question time, it was clear to me the audience members were mostly in a state of anxiety, frustration, even despair. They didn't need to be convinced, they needed to be inspired.

Thankfully, the last third of the presentation addressed possible solutions to climate change that were already in play or within reach, and you could see the crowd perk up and faces lift, if only a little. While the presentation lacked nothing in terms of academic rigour, it lacked a narrative that was emotionally satisfying, a recognition that what the audience needed was to leave feeling not necessarily cheerful, but resolved.

The one-way model of science communication is slowly but surely being replaced by a more interactive and social approach to educating people about science. This involves all parts of our society in a collective conversation about not just causes but solutions. The citizen science movement is gaining traction around the world, and is an important part of improving not just general scientific literacy but people's ability to connect what's happening in our natural world to their own lives. And wouldn't it be wonderful if every country could follow the examples of some enlightened countries, such as Italy, where climate change studies are now compulsory? In these places, all school-age children must spend time each week studying topics like ocean pollution, sustainable living and renewable resources.

But even if we could snap our fingers and mimic these countries overnight, dramatically improving the level of science education around the world and our understanding of climate science, it would not be enough. People have to be convinced to *care* about it, be motivated and be provided with opportunities to take action.

It's not just about education but persuasion. Reason *and* emotion. The right balance between all these elements.

It was exactly this balance that I wanted to talk about with Miranda Massie. Like me, she came to the climate change cause late in life. It's not as if the civil rights litigator, an accomplished lawyer and Ivy League graduate, didn't care about the environment at all. It was just that she was more concerned with others issues, such as affirmative action. 'Like many, many other people in the United States today, I pushed it down for a long time,' she told me from her desk in New York City. Unlike many climate activists I've interviewed, Massie can't remember the first time she heard about global warming, but she can remember that impulse to 'push it down'. 'I remember resisting thinking about it too deeply, realising how overwhelming it is, and that it would require a wholescale change in my life.'

When superstorm Hurricane Sandy hit the United States in October 2012, causing damage in 24 states, flooding streets, tunnels and subway lines, and cutting power in and around New York City, Massie could no longer relegate climate change to the sidelines of her worldview.

> When Sandy hit, I'd been thinking about climate for a while, or rather trying not to think about my own sharpening sense that the climate crisis demanded a shift in focus. The storm changed my feelings definitively and I was aware of it at the time.

As a response to these changed feelings, Massie came up with the idea of a climate-dedicated museum. The thought came to her in the weeks after Sandy and seemed so obvious that she assumed it was already up and running, and that all she needed to do was track down the director and offer them her services. 'I was astounded to learn that a climate museum wasn't already underway,' she recalled. And so Miranda decided to start the process of establishing one.

This change in her professional and personal focus towards climate change was, Miranda realised, to take on a significant and encompassing anxiety. 'You know you're going to be uneasy and uncertain indefinitely.' Now she lives and breathes climate change as the director of the Climate Museum. It's the first of its kind, with a mission to inspire action on the climate crisis through programming across the arts and sciences that deepens understanding, builds connections and advances solutions. While the offices are based in New York, the museum doesn't have a permanent home or dedicated space (although that's the ultimate aim). Instead, it stages exhibitions in different places around the city. Exhibitions so far have focused on a mix of information, art and inspiration—issues such as the new youth climate movement and the melting polar ice caps.

When we think of a museum, we imagine a place dedicated to the storage and display of objects and information of scientific or cultural interest. It's supposed to be predominantly educational, a place where you leave feeling smarter, a destination for droves of school students on excursions with worksheets to complete. The Climate Museum's exhibitions are consciously different, melding the artistic with the statistics, more interactive art gallery than static display of taxidermy.

While you might leave one of the museum's exhibitions feeling more knowledgeable about the science, the main aim is to get people emotionally engaged and, ideally, to push the issue of climate change into everyday conversations. Miranda describes the raison d'être of the museum as providing opportunities for physical, emotional and social learning. She argues that while it's important to convey the numbers and graphs of climate change science, most of us (especially those without scientific training) need context and story. We need to be able to relate climate change to both our community and peer group. This helps us overcome the challenging and abstract nature of the scientific data and understand the gravity of its meaning.

I was intrigued to know (given the museum is a relatively new venture) what Miranda had discovered about its impact on the people visiting. She understands that the museum's core audience is people who are concerned about climate change but lack confidence about the science. While they're interested in learning more about climate facts and figures, their real hunger is for greater guidance and examples of how to act, especially in collaboration with other people. They want to be engaged by a social, physical and emotional experience. She believes that this sense of being inspired to act, especially in the context of a broader group or community, is far more important than any intellectual gains in understanding the science of climate change.

One of the benefits of not having a permanent home but mounting exhibitions around and about in public places is that the museum can capture the attention of people who just wander into the space, rather than those seeking out a climate change experience. For example, in 2019 they mounted an exhibition by Justin Brice Guariglia at ten sites across the city's five boroughs, featuring solar-powered highway signs flashing prominent, unexpected messages about climate change in different languages. 'Fossil Fueling Inequality.' 'Climate Denial Kills.' 'Climate Change at Work.'

Some of the phrases were deliberately provocative. But they didn't just stand alone in the urban landscape. Part of the exhibition involved hosting speaking events and walking tours with social justice organisations, scientific researchers and local environmental advocacy groups to start conversations with visitors and passers-by.

The museum's next major effort was a show on Governors Island focused on carbon solutions and on taking civic action to get these solutions implemented. Again, they organised speaking events and tours, ensuring conversations were a part of the visitor experience. They had 27 high school docents leading visitors through the show. Miranda told me she was most excited about the accidental visitors and the ways they were encouraged to take action:

> A lot of people came through because they were walking around Governors Island and we just nabbed them. Those are our favourite people in some respects because they haven't yet been active but can be inspired to take a fresh approach because of a museum experience . . .

> The general population is ready to be moved into a climate-active stance if we can provide them with spaces and opportunities to do so with a sense of social safety.

The challenge is to strike the right balance between presenting the (overwhelming) scientific facts and inspiring the emotional and social impulse to act. Miranda believes there's no set formula for achieving this balance, given how diverse the audiences are and the rapid rate of climate change itself:

> With the crisis accelerating, there is a need for a strong dose of how catastrophic current impacts are. That's required in order not to insult the intelligence of the visitor and to be an honest broker. But you can't lead with that with most people and expect them to stay with you. You can provide a reasonable argument based on the science that we have a chance to stabilise the climate to a point where we can sustain our civilisation. If you do that, people can leave an exhibition not necessarily feeling cheerful or even hopeful, but with a real sense of their own capacity and resolve.

Miranda makes the critical point that while we might comprehend the science of climate change to a greater or lesser degree, the vast majority of us need a larger story, context and connection—both to the issues we care about and the social group we belong to—if we are to be motivated to actually do something with this knowledge. We need a personal framework for the scientific data, an emotional wrapper around the rational material. Indeed, that was what Miranda herself needed. What tipped her from being generally worried about climate change to being part of the growing ranks of climate advocates around the world was an emotional and social reaction to a storm that flooded the streets of her beloved New York City.

My aim in this chapter hasn't been to argue that facts don't matter or the scientific method should be watered down or that we should communicate without facts. I'm not suggesting that emotional appeals always work better than reasoned arguments—on the contrary, using the wrong emotion with the wrong audience can backfire badly, as you will see in the rest of the book.

What I am saying is that now that the climate science has been proven to be true to the highest degree possible, we have to stop being so reasonable and find more and more 'irrational' ways to talk about climate change.

CHAPTER 2

START BEING EMOTIONAL

Or the importance of feelings over facts

Tony Leiserowitz looks as much like a recovering ski bum as he does a world-renowned Ivy League professor. He is trim and neat, with the healthy glow of a man who enjoys the mountains as much as the wood-panelled offices of his workplace. I am sitting across from him in his study on the top floor of the Yale School of Forestry and Environmental Studies, where he heads the Program on Climate Change Communications. Since 2005 the program has been leading the world in research on how to communicate effectively about climate change.

I'd travelled to Yale on a kind of pilgrimage to spend a week in and around the program, talking to Leiserowitz and his many talented social science grad students and researchers working on various projects—everything from how Christians respond to climate to how voters feel about wind farms. Visiting the program was like a boot camp in understanding climate change communications.

Leiserowitz describes his path towards his area of expertise as 'winding'. He studied international relations at university, namely Cold War politics, with a view to a long career trying to prevent the United States and Russia from blowing each other up with nuclear weapons. But six months before he graduated, the Berlin Wall came down and, as he puts it, his international relations degree turned into a history degree overnight.

A passionate traveller, he followed a friend to Aspen in the hope of making enough money doing any kind of jo b around the resort town to fund some international travel. Instead, he lucked into a job at the Aspen Global Change Institute, where he spent four years learning about climate change and biodiversity extinctions from some of the world's best environmental scientists.

Leiserowitz arrived at the institute at a critical time in the history of our understanding of climate change. It was the early 1990s, a few years after the famous testimony of the astrophysicist James Hansen to a US Senate committee. In 1988 Hansen told a phalanx of politicians (including a young senator from Tennessee, Al Gore) that 'the greenhouse effect has been detected and it is changing our climate now'. We could almost say Hansen's testimony marks the beginning of the public's understanding of climate change. Certainly, it helped focus political and policymakers' attention on the issue, and much more grant money started to flow to natural scientists looking at the phenomenon.

Even though he was getting a world-class education in environmental science at the institute, Leiserowitz was feeling frustrated at the lack of a social science dimension to the work they were doing. In an interview for the podcast CleanCapital, he reflects on this frustration.

> The natural scientists [there] were fantastic and I loved being with them. But personally, I felt we were mostly talking about symptoms and not the underlying causes, because if you look at any global environmental challenges, the reason why they exist is human beings. The natural scientists are incredibly important for helping us understand these challenges and why they happen, but ultimately, they're the result of human perceptions, human decisions [and] human behaviour. I felt if I wanted to really address these issues, the answer was not going to be in the natural sciences. It was going to be in the social sciences and even the humanities.[1]

Looking back at the list of questions I asked Leiserowitz that June day in his study, knowing what I know now, they seem painfully unsophisticated. I basically went through all the emotions I could think of—fear, love, anger, guilt and so on—and asked him whether they worked or not when it came to communicating about climate change. His patient responses were a variation on a theme. It all depends.

Human behaviour is not a simple arrow from attitudes to behaviour; emotional appeal plus 'call to action' doesn't equal response. You have to start with an appreciation of human psychology but also with a nuanced understanding of the mindset of your intended audience. Before you even think about what to say, you have to ask yourself who you're talking to and what you want to get them to do. 'Context matters,' Leiserowitz explained. 'There's no simple formula: three parts hope for every one part fear.'

If only it were that easy. And given that the climate change message has been circulating in the public domain for decades and large groups of people still resist and ignore it, Leiserowitz told me that increasingly we have to ask ourselves not just who we're talking to but who those people want to listen to.

'The messenger is often more important than the message,' he said. That's a fundamental principle of communications but worth repeating time and again when it comes to an issue as politicised as climate change, where famous advocates can make people tune out as much as tune in. While climate change is, as Leiserowitz put it, 'the mother of all collective action problems', requiring us all to band together to address the issue, that doesn't mean there's one right way to convince different people to act. 'The assumption is that everyone has to walk the same path you did.'

One of the biggest tasks of the climate movement, Leiserowitz told me, is to create a positive, alternative view of the future. It's a task they haven't yet successfully completed, he argued.

> We've been much, much better at describing what we're against than what we're for. We paint a picture of hell for people, a place of fires and plagues and floods. We don't tell an alternative story of a better world we want to live in.

That leaves a cultural vacuum for our opponents to fill and they're happy to do it. Conservatives that oppose action on climate change love telling people that environmentalists want to take away your house and your car and leave you shivering in the caves. They get away with it because we don't have an alternative vision.

Creating an alternative vision that we all support and want to work towards requires a deep understanding of the hopes, aspirations, values and mindset of different groups of people. It requires identifying common ground while also understanding, respecting and working with the differences that exist in any society. One way Leiserowitz and his colleagues at Yale and George Mason universities try to do that is through their Six Americas study.[2] This large-scale survey measures the public's climate change beliefs, attitudes, risk perceptions, motivations, values, policy preferences, voting patterns and media consumption, and the underlying barriers to action. Depending on how different people respond to these issues, they break down or 'segment' into six different groups, with people in each segment sharing certain attitudes, values and even demographic characteristics.

At one end of the Six Americas spectrum we have the Alarmed, fully convinced of the reality and seriousness of climate change and already taking individual, consumer and political action to address it. (I am firmly in that group, and so are many of my friends and the people I work with.) The remaining groups vary in their degree of anxiety about climate change—in descending order the Concerned, the Cautious, the Disengaged, the Doubtful and finally the Dismissive. In societies like the United States, Britain and Australia, understanding attitudes to climate change according to these kinds of groupings gets us towards a deeper understanding of why people react to climate change the way they do.

When I first read the Six Americas research, I pictured in my mind a big family seated at the Christmas table, with an alarmed teenage girl seated across from her dismissive uncle arguing about the relative merits of Greta Thunberg, and various members of the family around them adding their two cents worth to the discussion. The concerned on the side of the alarmed. The doubtful siding most of the time with the dismissive. And the disengaged mother hoping they'll all stop arguing in time to help her clear the table and do the dishes.

In the last decade or so there has been more and more research and interest in climate change from the humanities and, as I mentioned in the previous chapter, more recognition that social scientists are essential contributors to the climate change cause. Better late than never I guess but, like Leiserowitz, I feel it was a genuine shame this focus on the *anthro* part of anthropomorphic climate change wasn't a major focus from the beginning.

In his book *Losing Earth: The decade we could have stopped climate change*, journalist Nathaniel Rich documents the wasted opportunity we had, in the decade between 1979 and 1989, to address the issue of climate change. During that time, Rich shows that 'the world's major powers came within several signatures of endorsing a binding framework to reduce carbon emissions—far closer than we've come since'.[3] He takes us behind the scenes at the various meetings, conferences, hearings and summits during this period, when policymakers and scientists came to grips with the alarming data and the possible government response.

While this was happening, a small band of philosophers, economists and social scientists—a group Rich calls the Fatalists—were 'busy conducting a vigorous debate . . . about whether a human solution to this human problem was even possible'.[4] The Fatalists didn't bother trying to figure out the emerging science and whether it could be believed; they took for granted the worst-case scenario sketched out by the most pessimistic of the natural scientists. Instead, they debated whether humans, when faced with the distinct possibility of their own extinction, were willing to prevent it.

> The Fatalists wondered whether greater awareness of the problem really would provoke a sensible response. Was the threat of a distant catastrophe sufficient to motivate change? If so, how much threat, and how much change? We worry about our children's future and our grandchildren's futures. But how much, precisely? . . . Enough to compromise our living standards? . . . And what degree of certainty was required if so? The question had to be asked not only of individuals but also of nations and corporations. How much value did we assign to the future?[5]

Rich writes that the Fatalists attended the major summits on climate during this critical decade, alongside the politicians and natural scientists, but when they expressed their views, they were met with polite smiles but little else. 'No one seemed to listen,' Rich writes. At the World Climate Conference in the spring of 1979, it was the physical scientists who dominated the meetings. When the Fatalists presented their arguments, the physical scientists would simply 'nod along, waiting for the opportunity to resume debating the relative influence of radiative transfer'.

> That was their field of expertise, after all—clouds, oceans, forests, the invisible world. And so it happened that the economists, philosophers, and political scientists came to feel that, no matter how forcefully they issued their warnings, they were becoming invisible themselves.[6]

Those prescient natural scientists were right about probable catastrophic climate change if we didn't reduce our CO_2 emissions, but they were lamentably wrong about what it was going to take to convince people to act. They didn't fully appreciate that climate change is a social, cultural and political phenomenon as much—if not more—than a scientific fact.

The science behind climate change has been proven correct to the highest degree of certainty the scientific method allows. But climate change is more than just the science. It's a social phenomenon. And the social dimensions of climate change can make the science look simple—the laws of physics are orderly and neat but, as I've said before, people are messy. When social researchers like me try to analyse how a person responds to climate change messages the way they do, we're measuring much, much more than just their comprehension (or not) of the climate science. We're analysing the way they see the world, their politics, values, cultural identity, even their gender identity. It wouldn't be a stretch to say we're measuring their psyche, their innermost self.

In his book *Why We Disagree About Climate Change*, British professor Mike Hulme argues that this is one of the reasons we argue so much about the issue. 'The sources of our disagreement with climate change lie deep within us, in our values and in our sense of identity and purpose,' he writes. 'They do not reside "out there", a result of our inability to grasp knowingly some ultimate physical reality.'[7]

It follows that in order to help resolve, even to some degree, the conflict and disagreement about climate change in the community, we need to understand those different belief systems and the emotional responses and social forces that shape them. And take them into account when we communicate about climate change and what should be done.

This is even more important given how politicised climate change has become, especially in countries like the United States and Australia. American research from Leiserowitz and his peers has shown that reactions to climate change as a topic were becoming increasingly polarised along partisan lines around the late 1990s. He argues that the climate change views of Democrats and Republicans were not significantly different until the Kyoto Protocol negotiations of 1997, when policymakers started to explore possible solutions to global warming.

In an article for the academic journal *Risk Analysis*, Leiserowitz showed that in 2003, when respondents were asked in surveys for their first reaction to the phrase 'global warming', only 7 per cent reacted with words like 'hoax' or 'scam'. By 2010, that had risen to 23 per cent.[8] There was a parallel trend in the United Kingdom: between 2003 and 2008, the belief that claims about climate change had been exaggerated almost doubled from 15 per cent to 29 per cent.[9]

The huge success and positive impact of Al Gore's first documentary, *An Inconvenient Truth* (2006), had the less than positive side effect of strongly associating the climate change issue with the progressive side of politics. Today, as Leiserowitz comments, climate change scepticism and even denial in the United States have become part of a cluster of beliefs (along with anti-abortion and anti-immigration) that are obvious markers of Republican allegiance.

The situation in my home country, Australia, is much the same, with countless surveys showing that conservative voters are less likely to believe climate change is mostly caused by humans and more likely to believe it has 'natural' causes. A 2017 study showed that 'Australians who identify with conservative political parties are far more likely than those aligned with other parties to reject anthropogenic climate change', with conservative men living in rural areas the most likely to be sceptical.[10]

In my own social research with Australian voters, I see this politicisation all the time. Nowadays, I don't even have to ask how someone voted in the last election to hazard a guess about their views on climate change. Sometimes all it takes is for me to ask them how they feel about the role of government (Are you taxed too much? Do you feel there is too much regulation?) and what media they trust the most (blogs and social media or public broadcasters?).

Of course, the media, including social media, plays a huge role here in heightening the politicisation of the issue and conflict over climate. In a 2019 paper for the US-based Climate Institute, researchers Isobel Gladston and Trevelyan Wing described and analysed the way social media has become a 'significant polarizing vehicle' when it comes to climate change.[11] While they believe social media can be a valuable platform for interaction between people of different views, at present it can encourage vicious cycles of opposition, creating echo chambers ruled by more extreme opinion leaders. It is certainly based on algorithms that reward outrage and amplify conflict, regardless of whether that comes from the right or the left.

The degree of polarisation over climate change in places like Australia and America is not universal across the globe. The esteemed Pew Research Center found in 2015 that in 'Canada, Germany and the United Kingdom, followers of conservative parties are much less likely than followers of liberal or green parties to believe they will be harmed by climate change'.[12] But in many other countries there are much less pronounced political differences on climate change, and much less public and political interest in contesting the science.

For environmental activists in these less polarised countries—often countries already feeling serious impacts from climate change but emitting negligible amounts of CO_2—the endless debate about the truth of the climate science in the big Western countries is gobsmacking. In the interviews I've conducted for this book, these activists have expressed their frustration and disbelief on this point. It's contributed not a little to their despair about progress at an international level.

As part of the research for this book, I interviewed a Philippines-based ocean conservationist, Anna Oposa, who reflected on this very point—how shocking it is that in countries like the United States and Australia, some political leaders are still arguing about the causes of climate change:

> When I first attended a United Nations climate change conference as a youth delegate, that's when I realised climate change is so political in other countries. I thought it was an advocacy issue, not a partisan issue. In the Philippines, when you talk about climate change, people get it. They get it's related to the storms and the flooding. I haven't come across anyone who is a true sceptic. You will meet people who think some of it is natural, but no one has ever said, 'Climate change isn't real, you're making

this up to get funding.' When I listen to the news from the US or Australia, I think, is this for real?

So when it comes to talking to people about climate change, it helps enormously to think about it not just as a scientific question but as a social and political one, increasingly so and more in some countries, and with some groups of people than others. But understanding how people's already existing (and often entrenched) political allegiances influence their response to climate change is only part of the picture. Understanding their emotional reactions is even more important, and that leads us away from the realm of politics towards psychology.

As I've delved deeper into reading about psychology, the various ways focus group participants react to climate change has made more and more sense to me. As I said before, I didn't study psychology at university but law and politics. I'm not one for regrets, but I wish I'd taken at least a few psychology subjects at the time. My understanding of psychology has been picked up over the years through some reading and absorbed over my fifteen years as a researcher conducting thousands of focus groups across Australia and reviewing similar social research from across the globe.

Viewing the climate change issue through a psychological lens yields endless important insights into why we are where we are. Have a look at the must-watch 2018 TED talk by American meteorologist J. Marshall Shepherd, '3 kinds of bias that shape your worldview'.[13] As a self-proclaimed weather geek, he often gets asked if he believes in climate change. He finds the question odd because science isn't about belief. It's about proof that things are real or not. He is agog at the chasm between what scientists know to be true and what surveys have shown the American public believes on issues like vaccinations, evolution and, of course, climate change.

This has led the natural scientist to start thinking about psychology, namely what biases shape our perceptions of the world around us. He picked three big ones. The first, and probably the most obvious, is *confirmation bias*, namely that we zero in on evidence that supports what we already believe. Confirmation bias is even more pronounced in a world where we can use our social media to filter out information we don't want to absorb and where we follow influencers who reinforce our already existing beliefs.

The second bias is called *Dunning-Kruger*, which describes our human tendency to think we know more than we do as well as to underestimate what we don't know. Again, I see this happen in focus groups all the time, when participants with no scientific credentials or training pick apart the science of climate change.

The third and final bias is *cognitive dissonance*. When people encounter actions or ideas they cannot reconcile psychologically with their own beliefs, they experience discomfort. They then try to resolve their discomfort by arguing away the new evidence until it's consistent with their own beliefs.

Given climate change is such a discomforting topic, I see this cognitive dissonance happen all the time in focus groups, where people try to find reasons other than climate change for the events happening around them, even when faced with a strong scientific explanation. They pick it apart because of Dunning-Kruger and then because of confirmation bias try to find a blog that states something other than what the scientific evidence shows.

J. Marshall Shepherd argues that we need to close the gap between public perception and scientific fact, in order to create a better future and preserve life as we know it. He challenges us to take an inventory of our biases and of the beliefs we use to prop them up. Think about where you get your information, how reliable it is and whether you only read the things that agree with what you want to think rather than the actual truth. Then share what you've learned—about yourself and about the world—with other people.

It's sage advice. We need to know our own biases before we can overcome anyone else's.

In the last two chapters I've tried to convince you that when it comes to climate change, we need to stop being reasonable and start being emotional. More science isn't the solution. *People are the solution*. Namely, understanding how they work as rational and emotional, messy and inconsistent, ingenious herd animals.

In this book I specifically examine emotions—everything from fear to love to guilt and despair—to understand how humans react to these emotions and what that means for how we can talk about climate change. Luckily, the specific roles emotions play in understanding how we respond to climate change are receiving more and more attention from researchers.

The relationship between emotions and climate change action is not always consistent, easy to measure or even predictable. But there's no doubt that a deeper understanding of our emotions is the key to finding a better way to talk about climate change and getting people to act—as the teenagers involved in the school climate strikes have shown.

CHAPTER 3

GREEN GIRLS

Or what we can learn from teens on talking about climate

Given they were instrumental in triggering my 'change of heart' on climate change, the 'green girls' of the climate movement were one of the first things I looked into when seeking to learn more about how we can talk about climate change in a way that makes a difference. What motivates them? How do they motivate others, me included? And what can we learn from them?

This brought me, one mild Sunday afternoon, to Daisy Jeffrey's house a few suburbs away from where I live in Sydney, Australia. After I greeted her parents and asked their permission to interview and record their daughter, Daisy and I sat at their kitchen table, cupping mugs of tea and picking at the rocky road I'd brought as a gift. Large, brown, intelligent eyes regarded me through bold spectacles. Daisy sports a glossy brown bob and has been blessed with perfect skin. She is cute and fierce at the same time. Sixteen years old, she is a student at the prestigious Conservatorium High School where she plays the cello (I assume much, much better than I did at her age). At this moment, she isn't particularly interested in a career in music. 'It's really tough to get a good job as a musician. There are so many amazing musicians who are in their late twenties in a whole lot of casual jobs and they're struggling to pay rent on a tiny flat. There are other things I'm passionate about.'

One of those passions is climate change. Daisy is one of the main organisers of the students' School Strike 4 Climate Australia. She has appeared on TV and on radio as the voice of the strike. Months after we met, Daisy, along with youth activists around the world including Greta Thunberg, stormed the stage at the climate talks in Madrid, frustrated at the snail's pace of progress from world leaders. I wanted to meet her to try to understand how she and others like her had managed to motivate so many of their peers to be involved in the strike.

Concern among the young, particularly girls, about climate is not surprising. In countries like Australia, the younger you are, the more concerned you are about climate change, and surveys show there is a gender gap, with girls and women expressing more concern about climate than their fathers and brothers. Daisy tells me she has grown up with an interest in environmental issues. At the age of seven she wrote a letter to the prime minister with an idea about how to recycle foil waste. A few years later she and some friends started a blog called 'The Environmentals'. Then for a few years she wasn't as active, although she would still have described herself as environmentally conscious to anyone who asked. Then the strikes happened. And things changed.

She attended the first strike as a participant and was overwhelmed. 'I thought, "This is different. These kids have had enough and they're using their voice, going out onto the streets." I was just so impressed.' One of the mums who volunteers at her school's cafeteria approached her after the strike and put her in contact with the main organisers. 'I've always been really outspoken at school. She thought this is something I'd be really interested in.'

For the second strike, Daisy managed to get 50 out of the 160 kids at her school to attend. She was a central figure in organising the second strike and then the third, making an impassioned speech in front of thousands of people. On one of her regular TV and radio appearances she debated a senior, climate-denying federal politician on the Rupert Murdoch–owned Sky News. She and some other climate kids went to the nation's capital, Canberra, to visit politicians and lobby them about policy. And she has received the inevitable hate mail via social media that comes with climate leadership.

What interests me most is how Daisy goes about convincing other students not just to be concerned about climate but to act on that concern. To do something their parents may or may not encourage, certainly something their school didn't actively encourage at the beginning, something that would have attracted significant criticism from people around them, including some powerful community leaders. To protest en masse in the way their grandparents perhaps did during the Vietnam War days.

In order to motivate people at her own school, Daisy made the usual pitch at the school's weekly assembly. 'I said there's a strike happening, this is why it's happening, these are the three demands, and I would really, really love if you could come and help show your support. And I did throw in the little bit about you'll be missing school.' There was some clapping from supportive people after she spoke, and about twenty of her fellow students signed up then and there. But the rest Daisy recruited through quieter, one-on-one conversations, something she thinks can be far more powerful than one message delivered from a microphone or, even worse, a screaming argument in the school hallways. 'I think shouting at people doesn't really help in getting anyone to go. I've learned that the hard way . . . People underestimate the power of just taking the time to sit down with one person. If you can even bring that person's opinion like halfway across, it makes a really big difference.'

It's not just schoolfriends Daisy can convince, it's strangers as well. One morning she was putting up posters and chalking the pavement in front of another high school in her area with the details of the strike when she overheard a conversation a few metres away between a father and his sons. 'He was saying "Oh, it's those bloody lefties with their protests."' Daisy decided to go over to talk to them. 'He was, like, so shocked that I approached him. I told him who was running the strike and described a couple of kids doing it. And I said, "You're by no means obliged to agree with me, but I just wanted to clear a couple of things up." And his sons came to the strike. The dad, I don't think he completely agreed with what we were doing, but there was a kind of understanding he got at the end of our talk.'

Daisy thinks there is always going to be 'peer resistance' to any political message that involves getting people to care. 'People will tell me, "I recycle, I think I make my individual difference." I'm like, "That's great, I'm happy that you do that," but I think people really underestimate the impact the strikes have in bringing climate change to the forefront of the discussion in Australia.'

The other barrier to action for schoolchildren, in her view, is the relentlessly negative image of the world projected by the media. 'We look at the media and we often don't see the good news, we just see all of the terrible eye-catching news. You see it happening around you and you kind of create a bubble around yourself. If you don't think about it, you don't process it, you don't raise any awareness about it. You kind of talk about it like it's off in the distance. It's like it's not there.'

Daisy describes herself as politically left-wing but she worries that climate change has become too partisan an issue. 'Climate change is very much everybody's issue, but it's kind of been painted as a left issue.' And unlike some of her peers who easily deride their opponents in the climate change debate or attack people working in mining, Daisy has more empathy than antipathy for those employed in these industries. 'My grandad was a coalmining engineer. When the industry tightened he was left without a job and ended up working as a cleaner. He'd been really passionate about being able to fix up these mines all over the world, and all of a sudden he'd just been stripped of his identity. His university degree was worth nothing.'

Listening to Daisy speak, it's clear she understands the power of empathy as much as anger to get the message across about climate change. 'I think one of the real faults in the climate activism community is that sometimes we forget to add faces to the people whose lives are built on mining. It's difficult to understand why people don't want to do the best by their kids and keep wanting to rely on coal, but what we've got to understand is that a lot of these people feel like they *are* doing the best by their kids and really think that coal is our best option.'

Given Daisy's preternatural capacity for empathy, it's not surprising her role model is the New Zealand prime minister Jacinda Ardern. 'She has a calm and compassionate way . . . She doesn't really yell about things. It's fine to be powerful and want to yell, but it's really important to be able to put that aside and just sit down and have a quiet conversation with someone. Teenage girls are good at that.'

While Daisy has helped drive the students' climate strikes in Australia, being involved in organising them has also helped her. Like so many teens around the world, especially girls, Daisy went through what she describes as 'a crappy mental health period' in her early teens, before her climate activism. 'I was struggling to find a point. I was like, "The sun's going to swallow the earth eventually, why even bother at all?" And I think the conclusion that I came to was, if you're going to be around, you may as well be happy and try to fight for what you believe in.'

After the 2019 federal election saw a conservative government returned to power in Australia, one that has a significant faction of climate deniers in its leadership group, it would have been easy for the climate strikers to despair. And there were tears and anger on election night, but interestingly not for long. 'We had this huge national group chat with everyone who's involved in organising both regional and major cities in Australia at about 11 o'clock on election night, and we just went, "Great, what are we doing now? Because we have to do something."'

Waving goodbye to Daisy and her parents on the street outside their home, I wondered whether her musical training and her skills as a climate organiser were related. When I asked her whether studying the cello had taught her anything applicable to the rest of her life, she responded this way.

> Music is like another language. Expressing yourself through your playing and doing regular performances at school has made me slightly less self-conscious about presentation. Being part of a massive group and playing music together kind of brings a sense of community and togetherness. I suppose it made me learn how to look at things from different angles, because you always have to interpret the music in your own way, but it also really helped me when it came to kind of making connections with people emotionally.

It's the ability of teen girls in particular to make emotional connections and talk persuasively to others that makes them such effective climate change campaigners. While Greta Thunberg is the most famous of all the young climate activists, reviewing video from around the world, it's clear that young women are front and centre of this movement. Most of the time, they're the ones speaking at the rallies and to the media. They are, as Daisy pointed out to me, really good at this.

There is research to back up the notion that teen girls are particularly persuasive when it comes to telling a climate change story. Academic Danielle Lawson and her colleagues from North Carolina State University wanted to understand the extent to which children and teens could influence their parents on climate change. Their starting theory was that intergenerational learning could be a useful pathway to building climate change concern within the community.[1]

In countries like America, Canada and Australia, one of the biggest factors influencing attitudes to climate change is political allegiance (as we saw in Chapter 2). Perhaps conversations between generations could bypass or cut through partisan divides. The researchers also theorised that perceptions about climate change in children are less susceptible to the influence of worldview or political context. Perhaps these emerging adults can be more open and flexible in their attitudes to politically divisive issues.

Lawson and her colleagues designed what they term an 'educational intervention' at a middle school (with students aged ten to fourteen) in a coastal town of North Carolina in the United States. They randomly assigned fifteen participating teachers to treatment and control groups of students from 238 families, trained them in a climate change curriculum specifically designed to promote intergenerational communication, and measured impacts on students and their parents over two years. They controlled for factors such as gender, race, political ideology and how much families talked about climate change, in order to better understand how these intergenerational conversations could operate across diverse family dynamics and demographics.

The results led to a number of media stories exclaiming that teen girls are the best communicators when it comes to convincing the unconvinced. The study made four major findings. First (and perhaps unsurprisingly), the children who were exposed to the climate curriculum became more concerned about climate change than the students in the control group. Second, those concerned students went on to talk about climate change with their parents. But the third finding was really surprising—the parents who were the most influenced by the concerned students were those typically resistant to climate change messages. As the authors commented, 'politically conservative parents who had the lowest concern levels before the intervention displayed the largest gains in climate change concern'.[2] Lastly—and this was the origin of the media headlines—the study showed that daughters were more effective than sons in getting their parents to be more engaged with climate change as an issue.

The last two findings are, for me, the most exciting. I'm intrigued that these intergenerational conversations could break through resistance to climate change messages, particularly with conservative fathers. In fact, the study showed that 'parents who identified as male or conservative more than *doubled* their concern levels between pre- and post-tests—a larger increase than their female and liberal counterparts'.[3]

The researchers also found that it was important that the information about climate change be as closely connected to local concerns as possible. The curriculum used by the teachers was hands-on, focused on local issues, and involved field-based experiences and encouragement of parental participation. Lawson and her colleagues made the climate change information personally relevant. 'As framing climate change in local contexts leads to increased climate change acceptance among sceptical audiences, local examples in the curriculum may have boosted learning among children and parents alike, including those sceptical of climate change.'[4]

And finally, the study showed that increasing family discussion of climate change was critical in predicting changes in parents' concern levels. Talking about it really mattered. Perhaps this is why daughters were more effective than sons in changing their parents' attitudes. Girls are generally better at verbal communication at that age (less grunty) than boys—as many parents who have tried to sustain a dinner-table conversation with daughters and sons about what happened during the day can confirm. Whether this is nature or nurture is probably irrelevant.

All in all, the research shows how including climate change in the school curriculum can be an effective way of educating the wider community on climate issues. It can also be the spark that creates climate activists; many of the men and women I interviewed for this book told me that learning about climate change at school helped them recognise the impact of global warming in their communities and was the catalyst for their development as passionate environmentalists.

This verbal dexterity among teenage girls combines nicely with the fact that in many countries some gender norms have shifted enough to give girls licence to be more politically opinionated, to take pride in their education and knowledge about the world, and to be more interested in STEM subjects. I wonder too—although there was no evidence to support this in the study—whether conservative fathers might feel a heightened sense of protection and concern for their daughters rather than their sons. They take the anxieties of the future mothers of their grandchildren personally.

Can this kind of intergenerational learning apply in other cultures, where respect for elders is more pronounced than in countries like America and Australia? Two academics from the University of the Chinese Academy of Sciences in Beijing found it can, but the other way around, with elders helping children understand the local impact of climate change. Sifan Hu and Jin Chen devised a new educational program for climate change in twelve rural areas in China. It included inviting tweens aged ten to thirteen to communicate in focus groups with seniors aged over 60.[5]

The aim of the focus groups was to discuss observable changes in the local climate during the past several decades. The status of older people in China was an important element in this study. 'Local seniors [are] trusted communicators ... perceived as "experts" with special status,' the authors stated. 'Moreover, revering elders and caring for the young is an important traditional virtue in China, which may increase the potency of this approach.'[6]

The older participants shared their memories of climate change (which were in line with actual scientific data). The influence of this on the tweens was significant.

> The adolescents' uncertainty about climate change exhibited significant change after the programme, followed by concern, risk perception, and perceived behavioural control. Based on mediation analysis, the shift in adolescent concern and perceived behavioural control translated into greater willingness to support climate change mitigation.[7]

In other words, the stories of these elders about the changing climate convinced the tweens, who then became more open to supporting action to deal with climate change. As with the North Carolina study, it was important that the focus was on local effects of climate change. This had the result of making 'the abstract concrete', 'the invisible process of climate change visible for engagement'.[8]

These two social experiments from two very different countries show the power of intergenerational discussions on climate change to shift even entrenched attitudes and to encourage action and support for climate change policies. The key here is to get people talking about climate change with the people they love and trust as an issue of immediate and local importance to their tribe. The messenger is more important than the message, as I've mentioned more than once. Today is easier to talk about than tomorrow. Local changes to the environment are more easily observed and more concerning than national or international ones.

Through the process of writing this book, I watched Greta-mania take hold in the media. In September 2019, Thunberg took a solar-powered yacht from London to New York to attend the UN Climate Summit, speak to world leaders and the world's media, lead the climate strike in New York and glare at Donald Trump. There were whispers about nominating her for the Nobel Peace Prize. Of course she has been criticised for her efforts by right-wing leaders in politics and the media around the world. Thunberg's appearance and the fact she has Asperger's, a condition on the autism spectrum, has been part of some of these attacks.

While denigrating her looks and her neurological make-up is both predictable and unconscionable, for me they are no doubt central to her appeal and effectiveness. She appears much younger than her actual age of sixteen. With her cherubic face free even from lip gloss and her Heidi-style plaits, you might think she was ten years old. There is her frank and fearless manner of speaking, unsullied by the usual teen speak ('Climate change is, like, a really bad problem. It's not awesome in any way . . .'). And finally, her relentless focus on the issue that matters to her most, her unblinking stare. When I watch her speak, I am reminded of all those times my own children have asked me questions I find difficult or uncomfortable to respond to because they touch a nerve, make me confront something I spend a lot of energy pushing to one side. 'Mummy, why do you shave your legs but Daddy doesn't?' 'Do pigs have feelings?' 'What happens when I die?'

In his book *Don't Even Think About It*, expert in climate communications George Marshall writes about the role of young girls in public persuasion campaigns. They have often been central to these, either as innocent interrogators or as victims. You may recall this famous World War I recruiting poster, featuring a man with a little girl on his knee and the tagline 'Daddy, what did *you* do in the Great War?'

It was meant to shame British men into volunteering to fight in order to avoid answering sticky questions from their progeny later in life. Only the original and real event that sparked the idea for this poster happened between a father and a son. The artist who painted the poster changed the questioner to a little girl, knowing instinctively the emotional punch would be greater.

George Marshall argues that this image cast the mould for many behaviour change campaigns since then that have involved shaming people into action as well as tapping into their love for their children.

> The moral dilemma that inspired this iconic poster is one of the recurring ethical themes in climate change

This famous recruitment poster was created for the British War Office in 1914. George Marshall cites it as the kind of behaviour change campaign that could be effective for climate advocates.

GREEN GIRLS 59

communications. Caring for the welfare of our children is one of our strongest evolutionary drives and one of the few concerns that consistently overcomes self-interest. On the face of it, giving those children a voice in our decisions, especially imagining how they might confront us in the future, should be a powerful spur to action.[9]

Marshall argues that while the poster might seem like social blackmail, it was in fact a highly effective combination of 'peer pressure, trusted communicators, social norms and in-group loyalty'.[10] He believes it's the kind of image and slogan that climate advocates should continue to be inspired by today.

It was a version of this scene that motivated me to become focused on climate change work. As my children grew up and realised the extent of the damage to the earth, I wanted to be able to look them in the eye and say, 'I know, I'm sorry, but also know that I was part of a movement trying to create a liveable world for you.' For me, putting my personal and professional efforts into climate change action sits in the category of 'things I do to protect my kids', like cutting up their grapes and warning them about stranger danger. But that's because I've already made the connection between concern for my kids and concern about climate change. And that's clearly not a uniform response.

In a post-election analysis I conducted for the NGO WWF Australia, we found that the more children in a household, the more likely the parents were to swing *away* from pro-environment parties.[11] Other studies in the United States, Canada and Britain have come to similar conclusions.[12] As Marshall points out, it's pretty easy and understandable that people with children can 'simply immerse themselves in the daily routine of tears, laughter, and the hunt for the missing shoe and put climate change into the category of tricky challenging things they would prefer not to talk about'.[13] So there are limits to the extent to which we can base environmental appeals on concern for our children. As Daisy herself remarked, people working for coal companies love their children too.

Greta is part of this tradition started with the wartime 'Daddy' poster, but also deviates in one important way. She is angry. She is pointing fingers. She is trying to shame us into action not through naive and innocent questioning, but clearly and directly. She has more than emotion and indignation on her side. She has science. And it's sending people wild in all different directions.

One speech she gave as part of her trip to New York—commonly described as 'angry and emotional'—triggered the greatest reaction from critics and trolls.

> This is all wrong. I shouldn't be standing here. I should be back in school on the other side of the ocean. Yet you all come to us young people for hope? How dare you! . . . You have stolen my dreams and my childhood with your empty words . . . You are failing us . . . But young people are starting to understand your betrayal. The eyes of all future generations are upon you, and if you choose to fail us, I say we will never forgive you. We will not let you get away

with this. Right here, right now is where we draw the line. The world is waking up, and change is coming whether you like it or not.[14]

Every right-wing pundit and politician must have wanted to ground Greta at this moment or send her to a military-style boarding school. Her content and tone resembles the righteous anger of a teenager who has just discovered her parents are hypocritical and full of it and has decided not to be their good little girl anymore. And it's hard to dismiss her because she has 99 per cent of scientists backing up her argument.

Of course she has her trenchant critics, as well as people who support the content of her message but wonder if she goes too far. She is the leading light of the movement but not always the right person to convince all groups. She herself realises that, consistently saying she is trying to get people to listen to the scientists. And she has inspired a generation of young men and women like Daisy who have—literally and figuratively—stepped onto the world stage to challenge an older, powerful generation to act quickly and decisively.

Young women have a unique skill set for being the climate change movement's most effective communicators. As we have seen, local and personal intergenerational communication is highly effective in convincing people not just that climate change is happening but that we need to act. But girls—and women—have an even more important part to play in the climate cause than as effective communicators. They can play a critical role in reducing global emissions, particularly in countries where there are already low levels of consumerism and consumption. And the more we do to provide them with real choices and economic and social power, the more we do for the environmental cause.

In her TED talk, writer and environmentalist Katharine Wilkinson argues that in order to bring down global emissions we need to give women and girls around the world the capacity to rise up. She believes that if we address gender equity, we can address global warming into the bargain.[15] We've known for some time that the more educated women are, the fewer children they have, and so educating women is critical to ensuring population control. If women seek an education and marry later, having fewer children adds up across the world and over time; Wilkinson quotes the statistics that 1 billion fewer people could help avoid 120 billion tonnes of emissions.

We also know that we need to change the way we farm and the way we eat; doing so will help with climate change mitigation and adaptation. Women make up the bulk of small farmers in the world, as well as being the main meal providers in households. Improving their productivity as farmers is an important part of ensuring food security. Wilkinson is involved in Project Drawdown, which estimates that improving small-hold women-based farming could result in a reduction of 2 billion tonnes of emissions by 2050.[16] Improving the living conditions of women and girls and addressing their needs in terms of better education and health, financial security, reproductive rights and agency at home could have a significant impact on addressing climate change. 'Women and girls aren't responsible for fixing everything,' Wilkinson says. 'Though we probably will.'[17]

Teenage girls are often criticised for being too emotional, a tendency we can attribute to nature, nurture, hormones or Snapchat. But it has turned out to work well for them when it comes to talking to others about climate change. They may not grasp the science in the way (some) adults do, but they understand the power of the precisely calibrated emotional appeal. They instinctively know that when it comes to climate, it's personal *and* it's emotional. And rational arguments based just on the science don't always work. In fact, science can be part of the problem when talking about climate change (as we saw in the previous two chapters).

So far I've covered the limits of science, examined the potential of emotion when it comes to talking about climate change, and looked at how the green girls of the climate movement manage to inspire the old and the young to take to the streets. It's now time for me to dive into the deep waters of the human mind and explore the full spectrum of our emotional response to climate change. Starting with guilt and ending in love.

An important caveat to the chapters to come. I've divided them into discrete emotional states and ordered them in a way that makes sense to me—and I hope will to you too. The progression of chapters is not, by the way, from negative to positive, good to bad. You'll soon see that all the emotions have a part to play, but each can also backfire under certain circumstances.

Just because I've concentrated on one emotion per chapter doesn't mean they're not all intimately connected. They are, in truth, very hard to isolate, and they work together in complex ways: fear and anger, hope and love, despair and fear, loss and love, and so on. Academic Daniel A. Chapman and his colleagues make the important point that rather than looking at emotions as a lever or switch we turn on and off when we want to persuade someone, we should focus on the nuanced and sometimes counterintuitive ways that emotion and communication are intertwined. A message about climate change that tries to inspire hope and resolve in one person may in fact provoke fear and anger in another, despair in their friend and indifference in their neighbour.[18]

I guess nothing worth doing is ever easy.

CHAPTER 4

GUILT

Or my plastic coffee cup killed the green sea turtle

It's 9.45 a.m. and I'm lining up at the cafe at work to get a coffee. I reach into my bag and realise I've forgotten to bring my re-usable coffee cup. Nuts. Is it in the car? No, I left it on the kitchen bench. I forgot it because I was distracted this morning. It was a crazy morning, also known as a typical morning.

I woke at 6 a.m., emptied and filled the dishwasher, ate breakfast, negotiated breakfast choices for my three kids, packed their lunches, made their beds, filled the dog's water bowl, hung up a load of laundry, checked social media quickly, negotiated clothing choices for my preschool-aged twins (it's like dressing Kardashians for the Met Ball), reminded everyone to brush their teeth, actually *brushed* one twin's teeth, brushed the twins' hair and argued about which hairclips to use, coached one of them through a meltdown about her socks (a sock crisis!), rushed them all out the door and into their father's car, ran around cleaning up mess, got in and out of the shower, got dressed, packed my briefcase, put the dog outside, checked the house was locked, got into the car, realised I'd left my phone inside, went back to get the phone and headed off to work.

Jesus, no wonder I need a coffee!

Still, I should have remembered my re-usable coffee cup. Do I have time to have the coffee here? No, I'm running late already. Bugger.

It's understandable that I forgot it because, as I have just illustrated, mornings are crazy. I get a lot done in the morning actually. (And I'm not alone. All over the world, women do most of the household heavy lifting, whether or not there's a man in the house. It's unfair that women do so much, but this gendered division of labour in the home exists across the globe, even in countries where women are well educated and in paid employment.)

But I should have remembered my cup. Because I know that Australians throw out 2.7 million single-use coffee cups every single day. This adds up to almost 1 billion coffee cups thrown out every year, a major contributor to litter on our streets and in our waterways. My one-use coffee cup will add to this avalanche of waste. It could fall out of the rubbish bin and into the gutter and drift out to sea and choke a turtle to death.

I'm about to spend $3.50 to kill an innocent turtle. As if they don't have enough to deal with already because of climate change. The sex of green sea turtles is determined by the temperature of the sand they lay their eggs in. Increasing temperatures as a result of climate change mean that more females are being born, disturbing the natural gender ratio. And that could mean, in the future, no more sea turtles.

I'm a bad person.

And my colleagues at work, they're going to see me walking around with a disposable coffee cup and they know all my research and writing is on climate change, and they're going to think I'm a hypocrite. Oh god, I hope that Craig guy who's obsessed with recycling isn't behind me in the queue . . . Okay, I'm going to get this takeaway coffee in a disposable cup *just this one time* (it's only one coffee cup in a billion) and from this point on I'm going to keep my re-usable cup in my bag at all times and this won't happen again.

And anyhow, single-use cups have nothing to do with climate change . . .

This *slightly* exaggerated description of my eco-guilt response to forgetting my re-usable coffee cup might be familiar to some of you. It comprises many of the thought processes that come with feelings of guilt, including negative feelings about self, anxiety about social rejection by the peer group ('cup shaming') and cognitive dissonance.

Guilt—and its cousin, the more extreme shame—are painful human emotions. But they're not always bad emotions. Professor Daniel Sznycer, a social psychologist at the University of Montreal, points out that both can compel us to improve our behaviour, especially towards others. 'When we act in a way we are not proud of, the brain broadcasts a signal that prompts us to alter our conduct.'[1] He argues that guilt and shame have played important roles in our evolutionary survival by ensuring we didn't harm people who cared about us. Guilt and shame helped maintain social cohesion, protecting the group dynamic so important to the survival of the herd. In hunter-gatherer societies for example, people relied on each other for survival. They had to band together, share resources and protect each other from threats. Not fitting in or not getting along with others could be a life-threatening move.

Guilt and shame, while both connected to real or imagined moral transgressions, are in fact different, with different triggers and effects. Scans have shown that they can light up different parts of the brain. Researchers who have studied the way guilt and shame activate different neural pathways have concluded that 'shame, with its broad cultural and social factors, is a more complex emotion; guilt, on the other hand, is linked only to a person's learned social standards'.[2]

What might the implications of this be for the role of guilt and shame in talking about climate change? Well, striking the difficult balance between the two can be important if your intention is not just to make the person feel bad but to get them to act, to change their behaviour in the short and long term. Psychologists often argue that guilt is the more constructive emotion, because it not only emphasises what someone did wrong but also encourages them to think about how to address the damage.

Given guilt is tied to beliefs about right and wrong, it can be used as a tool to overcome conflict and change behaviour. On the whole, psychologists see guilt as linked to what they call 'prosocial' behaviours, such as apologising or repairing damage through restitution. What's more, you can feel guilty but your belief in yourself as a good person can remain intact. In contrast, shame has been linked with 'antisocial' behaviours, in part because it makes us feel bad about ourselves as a person. If you shame someone in a way that makes them feel inferior, then you can provoke reactions like anger, withdrawal or denial of the issue.

Talking about climate change in a way that prompts us to feel a constructive level of guilt but not a destructive feeling of shame is tricky. As with everything, it all depends on your audience. But I often think about it in these terms. Effective appeals to act on climate change have to acknowledge that we all live compromised lives, that we all make 'bad' choices out of necessity or lack of options. And yet we still have a responsibility to care for the environment, change what we can and act as part of a larger group.

Constructive guilt emphasises collective responsibility as much as, if not more than, individual responsibility. 'We have the responsibility,' 'We can and we need to act.' Rather than 'You're wrong,' and 'You shouldn't have done that.' It draws some kind of a distinction between what we do and who we are. It can acknowledge goodwill and good intentions, even when we don't do the right thing all the time.

So making people feel guilty about climate change could work to get them to change their behaviour. Making them feel shame could easily cause them to back away and become defensive, pushing them further out of reach of the message. But there's another challenge facing anyone trying to use guilt in particular when communicating about climate change, one I just hinted at above. In order to feel these emotions, we first have to feel *responsible*. We have to feel as if we've caused the damage done.

The guilt I felt about forgetting my re-usable coffee cup is pretty straightforward. I know the contribution disposable cups make to my country's landfill problem and I made the mistake of forgetting to bring my re-usable cup. Instead of skipping the coffee, I put my selfish need for caffeine ahead of the welfare of the sea turtle. The result was guilt (and a little shame had I been caught by Craig).

Next time, I tell myself, I'll reflect on my mistake and (I hope) remember the cup, so the guilt response has resulted in (reinforced) behaviour change. But I also know that as a wealthy person living in a wealthy country (with dishwashers, cars, hot showers, domestic animals, multiple children's fashion choices, etc.), I'm contributing more than my fair share to CO_2 emissions just going through my morning routine. And yet it was the coffee cup and not my lifestyle that made me feel guilty that morning.

Psychologists interested in how we respond to climate change point out that the very nature of the causes of global warming works to sideline our moral judgements, making it even more difficult to get people to respond to triggers involving either guilt or shame. No one, not even the people working directly in the fossil fuel industries, wants climate change to occur. Certainly, every other working parent with a morning routine like mine is simply trying to make it to work in one piece rather than actively trying to increase global temperatures. Climate change is being caused by our behaviour, the result of all the energy we use to live our modern Western lives. But we perceive it as just a side effect, collateral damage. If we see it at all.

Looking at climate change as unintentional harm is important given humans tend to place huge moral significance on whether someone causes harm deliberately or not. Perhaps that's why we're so fond of saying to someone we've hurt, 'I didn't mean to hurt you,' as if the intention was the lion's share of the crime. Intentional acts provoke powerful emotions; unintentional acts less so. And with unintentional acts there's no villain, no individual to blame, just good people trying to do their best in a bad situation.

Despite the downsides of either guilt or shame, they have long been part of the emotional toolkit for climate change communicators. In *Don't Even Think About It*, George Marshall argues that this, in part, is a consequence of climate change being the main advocacy issue for green groups, so it comes with all the positive and negative baggage of the environmental movement. People who don't identify as environmentalists can sometimes view them as professional complainers, killjoys and worry warts, always focused on banning and stopping things.

There is almost a puritan ethos of disapproval hanging over environmental messaging. Everyone seems very *woke*. Stoknes comments that the condemnatory tone of a lot of climate communication has created strong associations with shame. 'Climate images and messages indirectly underscore that I should feel bad about the way I live.'[3] Deniers and those seeking to undermine the climate change movement know that painting environmentalists as grim foot soldiers of the nanny state is an effective strategy.

Another effective tactic is highlighting the CO_2-emitting behaviour of celebrity environmentalists. Al Gore has a private jet! Emma Thompson eats beef carpaccio! Leonardo DiCaprio has too many disposable girlfriends! Deniers understand enough about how humans respond to guilt and shame to know that letting us off the hook for our choices and attacking rich and famous greenies will be an easy win for them and a cheap thrill for us.

While I don't believe the characterisation of environmentalists as judgey zealots is fair, I understand why people respond the way they do to messages aimed at making them feel responsible for behaviour that harms the environment. I see it in focus groups every day. When some people in my research are confronted with the impact of human-caused climate change—especially those who are cautious and doubtful about the issue in general—their first impulse is to blame-shift. They minimise their own impact: 'My impact is nothing compared to the big corporations.' They justify inaction by emphasising the imagined inaction of others: 'I'm doing my best, but I know people in China aren't.' And they point out the things they *are* doing to show that they're in fact doing the right thing overall: 'I recycle, so I'm doing my bit for global warming.'

It's important to note that shame reactions to climate messages are more pronounced in wealthy countries, especially those contributing the most to climate change. If you live in Zambia or Kiribati, and you know your country contributes almost nothing to global CO_2 emissions, you have nothing to feel bad about. In contrast, people living in wealthy countries know that their much valued way of life is under scrutiny because of climate change. As Stoknes comments, 'for most people in rich countries, the climate message . . . is uncomfortable to live with' because 'it tells us that we are partly at fault for the destruction of the planet'.[4]

You can see this rich country/poor country divide play out in global surveys about climate change solutions. The 2015 Pew survey found that the greatest support for wealthier societies doing more to deal with climate change is found in relatively poor economies that are not major sources of emissions.[5] People in the Philippines (73 per cent), Ghana (64 per cent) and Tanzania (64 per cent) say rich countries should do more. Compare this to Japan, fifth on the global list of annual CO_2 emitters, where only 34 per cent of citizens believe rich countries should do more about climate change and 58 per cent say developing countries should do just as much as wealthy nations.

On the bright side, that Pew study also showed that a good majority of us around the world (67 per cent) believe that lifestyle change is required to address climate change, as opposed to believing the problem can be managed through technological solutions alone. While there were regional and country-based differences regarding support for lifestyle changes over improved technology, it's heartening to know that most of us understand we need to make changes to our lifestyles.

But a worrying aspect of this focus on lifestyle change works to increase specifically our shame response. Namely, that in affluent countries like Australia, action on climate change has narrowed down to what each individual can do around their own home or when they go shopping. A lot of emphasis has been placed on the domestic sphere, with the individual as the agent of change. Marshall explains why this has emerged as the focus for a lot of climate change messaging:

> According to 'self-perception theory', behaviours are an important cue for self-image. If someone can be persuaded to adopt environmental behaviours, [they] may over time come to identify [themselves] as someone with an environmental worldview . . . On this basis, in the early 2000s, environmental organisations began to focus increasingly on the personal responsibility of consumers for climate change . . . To bring it home, they distilled personal actions into lists of household hints.[6]

There is nothing wrong with encouraging people to be more aware of the chemicals, water and energy they use in their home or to get them to compost and recycle. But I sometimes feel as if climate change messages that address citizens *just* as household managers or consumers are too narrow to be effective. There is a limit to how much personal and household behaviour change can bring about a rapid move to a low-carbon economy without a policy framework and action from big corporations and the government.

This goes back to my point that constructive guilt emphasises collective responsibility whereas shame tends to point the angry, quivering finger at the individual. Sometimes I come across climate communications that look like they've been written by the love child of Martha Stewart and Al Gore, a long list of new and potentially time-consuming edicts about how I need to run my home, feed my family and shop for personal items. For those feeling stressed, exhausted, overwhelmed and powerless in their life *as it is now*, saving the planet seems to involve a lot of new products and more time spent on a home that already feels messy and expensive to run. And if I'm using my own research with Australians as a guide, people feel pretty ashamed about the state of their homes to start with.

So given what we know, particularly about shame but even about guilt, it's easy to see the risks involved when we use them in climate change messages. There can be backlash and we can reinforce perceptions of environmentalists as disapproving lifestyle fascists. But as I said, guilt has the capacity to be a constructive emotion. If we can successfully make people—especially wealthy people in wealthy countries—feel responsible for human-caused climate change and feel guilty about that, then we're closer to getting them to act. That's particularly important if they act in large numbers and through powerful organisations.

Indeed, while many psychologists like to draw a distinction between guilt and shame, others looking at these emotions in the context of climate action believe this division is not completely clear-cut. Three German researchers looked at whether the combination of guilt and shame—what they called 'guilty conscience'—could actually be effective in promoting not only pro-environmental attitudes but also pro-environmental behaviour.[7]

They recruited 114 German university students and divided them into two groups, presenting one group with information about human-caused environmental damage and the other with information about natural environmental damage. They were then asked for their emotional reaction, and what they were prepared to do in terms of action and changed behaviour in the future. Finally, a petition about environmental pollution was presented to them to sign if they felt so inclined (no pressure).

The German researchers found that the students who were confronted with the human-caused information reported significantly more guilty conscience (a mixture of guilt and shame but more towards the guilt end of the spectrum) than those people presented with the natural environmental damage. They also reported higher levels of positive behavioural intentions regarding the environment and were significantly more likely to spontaneously display actual pro-environmental behaviour like signing the petition. The researchers did own up to the fact, however, that university students might not offer the best possible general cross-section of the population and that the debate about guilt versus shame in environmental messaging thus needed further inquiry.

In contrast, a group of researchers from Columbia University found that a positive feeling like pride was far more effective than guilt in encouraging pro-environmental decision-making (the researchers made no distinction between guilt and shame in their study).[8] They recruited a diverse sample of more than 500 Americans, then randomly presented them with different environmental choice scenarios and asked them to make a decision between a 'green' or 'brown' (i.e. standard) option. Participants were asked to reflect upon either the future pride they might feel as a result of taking a particular pro-environmental action or the future guilt they might feel as a result of not taking a particular pro-environmental action or taking a bad environmental action.

The researchers found a positive correlation between pride and choosing the green option, and a negative relationship between guilt and the green option. In other words, anticipating a feeling of pride in the future because you made a good environmental choice was more motivating than the anticipation of guilt at not making the right choice. In this context at least, carrot beats stick.

In another study, research by psychologists from Cornell University showed that guilt doesn't always make us turn inward and reflect. It can make us lash out. In their survey of 700 Americans, Hang Lu and Jonathon P. Schuldt found that guilt may lead people not only to hold themselves responsible for negative outcomes they cause, but also to attempt to purge their guilt and responsibility by scapegoating others.[9]

Reviewing the writing and studies on the positive or negative role of guilt and climate change, the prevailing view is that trying to make people feel guilty and therefore responsible can be a risky strategy. It can lead to anger, resistance and avoidance, and can discourage deep listening and understanding. In their study on pride and guilt, the Columbia researchers concluded that messages based on negative frames like guilt may constitute a threat to a person's self-image and morality, and may lead to defensive reactions instead of the intended positive behavioural outcomes.[10] If the aim of the climate change movement is to persuade people to act, especially those people not already fully convinced and on board, then guilt may not be the most useful emotion to employ. It makes sense. You don't really want to listen to someone who makes you feel inferior.

But that doesn't mean guilt (with a small side order of shame) isn't useful in some circumstances with some groups of people. Guilt works on me. It works to keep me motivated to act on climate change. It was guilt—about the world my children will inherit—that brought about the transformation I described at the beginning of this book. It keeps me in the alarmed group of citizens, with action on climate change a motivating factor in everything I do. And when I think about the most persuasive, moving climate change message I've ever heard, well, guilt was a big part of why it was persuasive.

I mentioned earlier that I attended the 2019 Climate Reality training in Brisbane. Over the three days we heard speakers from around Australia and around the world—activists, politicians and scientists in the main. There were many First Nations leaders from different parts of Australia as well, including a group of local government representatives from the Torres Strait Islands. This cluster of around 274 islands lies between the far north of the Australian coast and Papua New Guinea. Around seventeen of these islands are home to about 6800 people (the majority of Torres Strait Islanders, around 42,000, live on mainland Australia).

Obviously, Torres Strait Islander communities have a unique culture and a long history with the islands and nearby coastlines. They can trace this history back over 60,000 years or more. The same problems impacting Pacific Island nations are affecting the Torres Strait Islands as well. Rising sea levels are destroying recently built sea walls, and extreme weather events are destroying infrastructure including houses and roads. 'Our island is being eaten,' said one resident interviewed by *The Guardian* about the deterioration of his home from rising water and strong winds.[11]

Before the Climate Reality get-together, I had read a bit about what was happening in Torres Strait, but at the conference it was the words of the Torres Strait Islanders themselves that affected me profoundly. On the first day, Mayor Fred Gela spoke frankly about the impact of climate change on the islands. He talked not only of the destruction of homes and the loss of land, but of how his community had been forced to relocate the graves of its elders and rebury them elsewhere for protection, without the appropriate cultural protocols. He reflected on the huge emotional and cultural damage this had caused to his people.

He spoke too about the fact that they knew they faced the prospect of relocation not only to different parts of the islands but for some to the mainland. 'That is our land. That is who we are ... Will we be the first climate change refugees in a country we belong to?' The next day, Torres Webb, an educator from Darnley/Erub Island in Torres Strait, showed us a photo of fallen trees being swept out to sea. He explained that when a child is born in the Torres Strait, a tree is often planted to mark their birth. These two events—birth trees being destroyed by rising tides, and the graves of revered elders being moved to protect them from being washed away—linked up in my mind. I couldn't think of a more powerful metaphor for the social, economic, cultural and spiritual impact of climate change on a community.

After his talk, I tracked Mayor Gela down outside the conference hall to thank him for his words and to discuss the prospects of the case a group of Torres Strait Islanders have taken to the UN Human Rights Committee (a world first) to pressure the Australian government into reducing national emissions. I wondered too, given the degree of destruction on the islands, why the Australian public wasn't paying more attention to what was happening. He sipped his coffee, smiled at me and said, 'Perhaps some people forget we are Australians too.'

More than 4000 kilometres (2500 miles) away from the northern coastline of Australia and from the imperilled Torres Strait lies the nation of Fiji. Whereas few mainland Aussies have ever visited the Torres Strait, Fiji is one of the most popular destinations for Australians when it comes to family holidays. They can arrive in Fiji, head from the airport to a gated resort and spend almost all of their time with other Australian families. I would hazard a guess that Australians would be more aware of the impact of climate change on Pacific Island nations like Fiji than on the Torres Strait. But Fijians and Torres Strait Islanders have something in common—namely, anger and frustration at the lack of the Australian government's action on climate change.

Whenever Australia's leaders get together with the leaders of the twelve Pacific Island nations, climate change is on the agenda and Australia ignores the unequivocal requests for action from these nations. All of these countries are suffering because of climate impacts, despite the fact that their combined contribution to CO_2 levels is a fraction of what rich Westernised nations produce.

Fijian climate activist Lavetanalagi Seru knows that when it comes to talking about climate change, guilt works. But you have to know your audience. Before Lavetanalagi learned about the science of climate change at school, he was experiencing it in his childhood village of Nanukuloa, in the Province of Ra, a coastal community on the northern part of the main island, Viti Levu.

Due to sea level rises, the people of his village built a sea wall to protect their homes. The wall held for a few years but soon it started to erode, the rising water undermining the foundations and claiming land on the other side. Then, in 2016, Tropical Cyclone Winston hit Fiji, causing widespread devastation to the economy and agriculture. Forty-four people died. Lavetanalagi's village was hit hard, with two-thirds of all homes destroyed or damaged. It was an event that helped the young man connect the science he had learned in school with the survival of his family and community. 'I thought, how many more cyclones can my village experience and still survive?'

From that point on, he started blogging about the climate and was active on social media. His Twitter feed is full of images of destruction caused by climate change. He helped found the organisation Alliance for Future Generations, an NGO championing the voice and meaningful participation of young people in sustainable development. And he started writing poems, like this one.

> *I stand, looking out to the ocean*
> *the land I stood on*
> *once, my childhood playing field*
> *Now engulfed by the sea . . .*
>
> *Inland, the sea water seeps*
> *into our family well*
> *Contaminating our drinking water*
> *from where my forefathers once drank.*

Now, left without a choice, we drink salty sea water . . .
Loss of land and a bleak future
Caused by deforestation and excessive pollution by humans
at whose hands, also lie the cure.

I asked Lavetanalagi whether he thinks guilt works to motivate people to act. He says it all depends. There's a big expatriate community in Fiji, including Australians and other wealthy Westerners, and Lavetanalagi thought his advocacy might convince them to become active in combating climate change. But he instead was met with indifference and condescension.

> An ex-pat came to me and said, 'Why are you blaming large industrialised countries for climate change? We've brought development to this country, and without us you wouldn't have developed as a country.' Another told me how costly it is for wealthy countries to move away from fossil fuels and 'It would be cheaper for those countries to pay to relocate your communities.' He didn't seem to understand, you are moving people not objects. These places are not just places we live, they are part of our identity, with deep cultural significance beyond monetary value.

That being said, Lavetanalagi and his organisation have had more luck with visiting groups of students and policymakers from countries like Australia, the United States and Europe who are already interested in the climate change issue.

We take them to communities impacted by climate change, to meet the people whose lives are being turned upside down. Yes, we're trying to get them to feel guilty, but we're also asking them, 'What are you going to do with that guilt when you return to your country? How can you make those leaders accountable?'

Lavetanalagi talked to me at length about the resilience of his community and the determination they have to continue to hold on to their lands, a huge task given many people have had to relocate to higher ground, abandoning places they have lived since time immemorial.[12] And yet he does struggle with feelings of resentment and anger that the world doesn't care about their plight. This is matched by a growing anxiety in him and other young Fijians about what the future holds if there continues to be a lack of international progress on climate change. 'I don't intend to have a child. I would love to, but I can't imagine the future for a child growing up given what's happening. Who wants to bring up a child on a drowning, burning planet?'

Perhaps what Lavetanalagi is trying to do with his overseas visitor tours is generate some compassion. Just as Fred Gela and Torres Webb are by talking about the Torres Strait Islands to conferences full of mainlanders. A sense of compassion, which should generate a sense of responsibility. The Cornell researchers Lu and Schuldt, in addition to their research on the limitations of guilt, have looked at the role of compassion and how it can be effective when it comes to talking about climate change.

That's because compassion has the potential to enhance social engagement with others and reduce psychological distance. This is critical because of our tendency to regard climate change in psychologically distant terms (far away in time and space).

Lu and Schuldt studied 400 American adults and showed that generating this compassionate impulse could improve the efficacy of climate change messages. The study divided the participants into two groups, both of which were shown an article about climate-change-induced drought in East Africa, with an image of a starving child in its mother's arms. The people in one group were asked to read the article and 'try to imagine how the child feels';[13] those in the other were asked to read the article with detachment and objectivity. The researchers thought they might elicit, respectively, a high-compassion and low-compassion response in the two groups. They then followed up with a range of questions about climate change.

The high-compassion group responded more favourably to support for government actions to address climate change. More importantly, given the known partisan divide between conservative and liberal responses to climate change, 'the high-compassion condition was more effective than the low-compassion condition in increasing support for government actions among moderates and conservatives'.[14] Generating compassion has the potential, the researchers argued, to break down partisan divides.

The researchers, however, pointed out one important caveat. Namely, that the climate victims portrayed in the news article were black and poor, living in another country. 'This may have elicited a great deal of psychological distance among our predominantly adult, white, American sample.'[15] Perhaps too much dissimilarity for compassion to overcome? It's a sad truth that we humans in general feel less compassion for those who don't look like us, even when they're fellow citizens and neighbours.

What, ultimately, can we say about guilt and shame in the context of climate change? Well, they work on me. Guilt about my kids and the climate kids converted me from concerned to alarmed almost overnight. Guilt and shame—about what's happening in the Torres Strait and places like Fiji—keep me engaged with climate change as an issue.

But as you can see, the psychological studies don't give us as much clarity as we'd like concerning the impact of either guilt or shame and whether we can distinguish between the two. Guilt can, under certain conditions with certain groups, help in behaviour change campaigns, keeping people vigilant about new social norms of behaviour and ultimately more supportive of government action. But shame is a riskier emotion to throw into the mix, especially if we make people feel ashamed about who they are rather than what they're doing. (Making coalminers feel bad about their jobs isn't going to stop coal production.) Overall, the work of academic researchers shows that positive emotions—like pride and compassion—may be more effective, especially with those who are just concerned about climate change rather than fully alarmed.

With so much guilt and shame sloshing around the climate change discussion, I suspect the people least likely to feel any are those who benefit from climate change the most. Does the CEO of ExxonMobil lay his head down on the pillow each night and think, 'I'd better bring my re-usable coffee cup to work tomorrow?'

I doubt it.

CHAPTER 5

FEAR

Or do wildfires change minds and votes?

In preparation for the Climate Reality training I bought a stack of books about climate change, among them American columnist David Wallace-Wells' *The Uninhabitable Earth*. It sat between Bruno Latour's *Down to Earth* and Edward O. Wilson's *Half-Earth* on my bedside table for the entirety of 2019. I was, in truth, scared to read it. I thought I would become overwhelmed with anxiety and fear if I did and it would affect my ability to put all my energy into my climate change work. A friend of mine had tried to read it and got about a third of the way through. 'More like The Uncontrollable Sob,' she told me. Somewhere in the middle of the year I opened it up and reviewed the contents. Chapter titles include 'Heat death', 'Hunger', 'Drowning', 'Wildfire', 'Dying oceans', 'Unbreathable air' and 'Economic collapse'. I turned to the first page and the first line. 'It is worse, much worse, than you think.'[1] I slammed the book shut. I needed to be a lot calmer and a lot stronger to keep going.

And yet I found myself finally reading the book during the Australian summer that saw unprecedented bushfires rage through most of the country, destroying hundreds of homes, killing 33 people and more than a billion animals. Armed forces were required to evacuate thousands of people, the biggest peacetime evacuation in our history. An astonishing 6 million hectares (15 million acres) of land burned up after the fires started in September 2019. Each evening before bed I would read in Wallace-Wells' book about a near future of millions of climate-related deaths, where large swathes of the earth are deserted, where countless major cities lie underwater, and where human life that does exist is grim and chaotic. In the morning I would wake to watch new details of devastation in towns I had visited for holidays and for work. My social media feed was full of images posted by friends of burnt-out houses belonging to friends and relatives. Over the months between November and January, the smoke from the fires blew into Sydney, Canberra and Melbourne, making the sun red and the air thick. Ash floated onto ocean beaches and in waterways.

Reading the book throughout this catastrophe provoked some wild dreams. I dreamt emus had invaded the house. I didn't want to evict them because I knew they would die of thirst and heat if I did. But they were angry and kept trying to peck at my daughters with their lethal beaks. I had to defend the girls with a broom. Reading it also induced some serious moments of depression and inertia, as I lay in bed contemplating the scale of the horror the author describes. *The Uninhabitable Earth* may be the perfect example of what Stoknes describes as 'collapse porn' (just a tip, whatever you do, don't google this term).[2] It's the kind of writing about climate change that seems to revel in all the details of the worst possible scenario. This sort of storytelling can bring about 'apocalypse fatigue' even in already concerned citizens like me.

In fairness to Wallace-Wells, his depiction of the near future is based on both scientific evidence and evidence of what is already happening today. I collect as many possible good news stories as I can about how we have the tools and technology to do something about climate change. But his book remains on my shelf as a reminder of what's at stake, what could easily happen if we continue on our current course.

As I mentioned earlier in the book, the bad news coming out of the natural science community about the progression of climate change is couched in academic language and so the full force of its impact on our way of life is often lost on the lay person. Scientists under pressure and scrutiny about their research are loath to speak in emotionally charged terms lest they invite more ridicule and aggression.

The aim of books like *The Uninhabitable Earth* is to paint a picture that's horrifying—the author wants us to *freak out*— as a necessary precursor to both accepting responsibility and doing something about it. Interestingly, Wallace-Wells describes himself as an optimist. He argues that given humans have shown themselves capable of changing the very atmosphere of a whole planet in such a short period of time, it's possible we may conjure new solutions, not to wholly reverse those effects, but to create a liveable environment rather than an apocalyptic one.[3]

Fear—like guilt—can work to keep me focused on climate activism. Fear is honest. Fear is more than justifiable. But is it effective? Does it help persuade people that climate change is indeed a crisis that requires our immediate response?

It's worth noting that our climate and extreme weather events like storms, droughts and floods have always provoked wonder and fear among human beings. British academic Mike Hulme writes that for much of human history, weather was 'beyond human understanding or control' and so appeared to mere mortals as 'the territory within which both divine and satanic influences were at work'.[4] Religions, both ancient and modern, have often viewed weather as a manifestation of the supreme deity's (or deities') approval or disapproval, with natural elements—fire, water, air—as weapons of the gods, to be used to punish or reward human behaviour.[5] Jupiter's thunderbolt, Poseidon's storms and floods, the great flood in the Old Testament that only the devout Noah, his family and menagerie survived. 'We love our climate,' writes Hulme, 'and yet we fear it.'[6] This might be particularly the case in Australia, where we have a climate of extremes across a vast continent.

This longstanding, deep relationship between fear and the weather should mean it's relatively easy to make people fear climate change. But that's not actually the case. Anyone with a basic understanding of how humans respond to threat and risk will point out that climate change poses a unique challenge. In order for a human to feel fear, we have to observe and register a threat, such as the sight of a predator or a situation that we recognise as dangerous to property, life and limb. That will then activate our 'fight or flight' response. While survey after survey shows that people around the world describe climate change as a serious threat demanding immediate action, we often don't behave as if this is the case. Climate change feels nebulous, and as such it fails to push all those critical evolutionary and cognitive buttons in our brains that are there to make us spring into action, to protect us against harm. As Stoknes points out:

> The climate crisis . . . is about abstract, imperceptible and gradual changes in weather trends from decade to decade. It is anonymous and not personified. It is beyond anyone's control and reach. It is rarely talked about at social events at the in-group level. It has a complex indirect impact on primarily far-off strangers, not us and our group. It is old and yesterday's news. Finally, there is no real enemy. If there is an enemy, it is none other than ourselves.[7]

Furthermore, he argues that the very fact that climate change is caused by CO_2, a colourless, odourless gas, makes this threat seem even more intangible.[8] Compare the build-up of CO_2 in the atmosphere with something like damage to the ozone layer of the earth, which dominated environmental concerns in the 1980s. The so-called 'hole' in the ozone (it's more accurate to describe it as depletion) was being caused mainly by human use of CFC gases in products like aerosol cans.

As Nathaniel Rich points out, 'the ozone hole alarmed the public because, though it was no more visible than global warming, ordinary people could be made to see it'.[9] I was in my late teens and early twenties when discussions about the ozone layer were at their apex. I can recall looking up into the sky on a regular basis, imagining a tear in the protective layers around the earth, letting radiation in like a weapon being used by some invading extraterrestrial force. It was easy to ditch the underarm deodorant spray as a result.

And so, as Rich writes, 'An abstract, atmospheric problem had been reduced to the size of the human imagination . . . been made just small enough, and just large enough, to break through.'[10] There was international cooperation to ban CFCs, and now the ozone layer does not face the kind of dangerous depletion it did a few decades ago.

Of course, if you care to look around you there are tangible, personal, imminent manifestations of the rapid build-up of CO_2 everywhere. Not just in sea level rises but in the form of extreme and unprecedented weather events. But you have to make the connection—that it's climate change created by humans driving these changes—to see the events as threats rather than just Mother Nature doing her thing.

Evolutionary psychology gives us some real insight into why it's hard to generate the necessary levels of fear about climate change to get people to respond 'rationally' to the threat.

Evolutionary psychology posits that human beings have evolved via natural selection over millennia in particular ways to survive in their surroundings. Our ancestors lived as hunter-gatherers and nomads and for thousands of years in small bands of twenty to 150 people. This way of life shaped what is described as the 'old mind' or 'ancient brain', ruled by 'ancestral forces': self-interest, status, social imitation, short-termism and risk vividness.[11]

It's easy to see how self-interest and short-termism being hardwired into the way we think can create challenges for a human response to something like climate change. But for the purposes of this chapter, let's concentrate on risk vividness. There are a lot of fancy academic definitions of 'risk vividness' but it boils down to this. If it's in my face, I'll be concerned about it. If not, *meh*. Anything that undermines that sense of the risk being real and imminent makes us less concerned.

What's more, as we saw in the early chapters of the book, humans evaluate the world around them with a mix of reason and emotion. Very few of us take a highly rational approach to the assessment of danger and risk. Stoknes puts it this way: 'psychologically, risks are feelings, not numbers'.[12] And there are different forces driving these feelings about risk.

First of all, as the evolutionary psychologists would point out, due to the ancestral forces of self-interest and 'risk vividness', we feel personal risks more than general or anonymous ones.

Second, apart perhaps from viral pandemics, we're generally less afraid of risks that are natural than those that are artificial or human-made. For example, people tend to be more afraid of radiation from low-level nuclear waste than from a few hours at the beach without sunscreen, even though the latter can be far more damaging. We also focus more on 'spectacular but rare risks' like a terrorist attack than on common ones like being killed in a car accident.[13] Related to this, we tend to focus more on risks that are being talked about in our social circles, by our leaders and the media rather than those that are less talked about, even if they're more objectively credible. So if it's in the public spotlight, we register it more as a real threat.[14]

Finally, we all have different attitudes to the notion of risk and therefore we respond differently to threats. There are risk-takers and there are those who are risk-averse. All kinds of psychological, social and cultural forces shape risk-averse and risk-taking personas. In my own research I've found that the people who are most blasé about the risks posed by climate change are young men. I recall in one project I was conducting about attitudes to climate change, the young men in the groups seemed entirely nonplussed about the prospect of environmental and social collapse. One of them remarked, 'If we're headed for a *Mad Max*–style society, well me and my friends are going to be okay. We're physically strong and we drive like crazy people.' On top of this, the fact that we have now had bad news about climate change for about three decades means, for some, the novelty has worn off, dampening their perception of risk. 'It hasn't killed us yet. Habituation has set in.'[15]

Some researchers have pointed out that the way we receive information about climate change has increased this distancing effect as well as made it less personally relevant. This has reduced the sense of threat and created further obstacles to generating the necessary level of fear. For example, for many, many years the pre-eminent image of climate change was the emaciated polar bear on a shrinking icefloe. It still manages to appear today in news reports about the impacts of global warming. Is it heartbreaking? Yes. Is it relatable? Unlikely (unless you study polar bears for a living).

Writing on how images of extreme weather symbolise human responses to climate change, British academics Brigitte Nerlich and Rusi Jaspal showed that media reporting can increase the distancing effect, making climate change feel less relevant and imminent.[16] They analysed the images the media chose to accompany reporting of the 2011 IPCC draft report on extreme weather and climate change adaptation. They found different kinds or groupings of images, but a common one was landscapes representing a deteriorating earth but devoid of humans. The absence of human beings in this context works to potentially increase our sense of distance between us and the environment, making the threat of environmental degradation seem less intense.

When humans were depicted, Nerlich and Jaspal found that people in the developing world were presented very differently from people in affluent countries. People in Bangladesh, for example, were presented as just 'getting on with life' in the face of catastrophic weather events associated with climate change. There was no sense of despair, of life grinding to a halt, but rather that this disruption had become part of their daily routine.

This has the potential to make people in affluent countries feel both as if climate change will only really impact people in other parts of the world and (even worse) that they will be able to cope because their living standards are so low anyway. In contrast to this group of images were those featuring people in affluent countries, where the focus was on devastation and loss—such as people crying in each other's arms as they survey the wreckage of a house destroyed by floods or wildfire.

The authors concluded that the images depicting climate change upon the release of the IPCC report emphasised fear, helplessness and vulnerability, which could be perceived as more passive than active emotions, and disconnected from activities associated with 'engagement and responsibility'.

But surely, once the extreme weather events and sea level rises increase, as predicted by the scientists, attitudes and actions will shift dramatically rather than incrementally. As I scanned the media, social and otherwise, during the Great Australian Summer of Smoke, the consensus seemed to be that given this was such an unprecedented event, it would amount to a tipping point in terms of attitudes to climate change, translating a general worry into a deep and persistent understanding that this was a national crisis and pressing emergency. Part of me wanted to believe this was true, but the social researcher in me was hesitant.

I knew that the research on whether extreme or unusual weather increases people's concern about climate change is decidedly mixed. A 2014 study by researchers from Columbia University found that ups and downs in the weather did in fact prompt concern about climate change.[17] However, perhaps due to the emphasis on 'warming' as an obvious sign of climate change, while unusually high temperatures tended to make people more convinced and concerned, during unusually cold periods, people's views went in the opposite direction. (That's why, all things considered, climate change is a better descriptor of what's happening than global warming, as it lends more emphasis to the transformation of weather patterns, including extreme winters.)

In my own work with Australians I've seen the same kind of response as the Columbia researchers documented. Weather events and hot summers bring climate change into everyday conversations around water coolers and dinner tables. And then the focus wanes as the weather gets cooler and a new 'crisis' emerges to attract media and political attention, such as an economic downturn, a pandemic or a terrorist attack.[18]

Similar research has found that temperature rises don't necessarily drive climate concern as much as people assume it might. Academics Parrish Bergquist (formerly at MIT and now at Yale) and Christopher Warshaw (from George Washington University) reviewed data on climate change opinions over the period 1999–2017 to see if public concern correlated with temperature rises.[19] They found that higher temperatures did lead to greater concern. Specifically, a 1 degree Celsius increase in temperature led to an increase of approximately 1 per cent in the share of people worrying a 'great deal or fair amount' about climate change.[20] While the relationship between warming and concern persisted even in the face of growing political polarisation, the problem of course is that the increase over this period was small. The two academics concluded that 'a warming climate is unlikely to yield a public consensus about climate change'.[21]

Our belief in and concern about climate change may rise and fall with the barometer and increase over time as more and more 'once in a lifetime' storms, fires and floods occur, but given the pace thus far of climate driving public worry, we can't just rely on that. Over time as well, these extreme weather events might become more familiar, more expected and then less and less effective as 'teachable moments' about climate change. Wildfires will become the new normal. Or, as Nathaniel Rich summarises it, 'disasters alone will not revolutionize public opinion in the remaining time allotted to us'.[22]

All this doesn't mean we should completely shelve fear as a useful emotion when it comes to talking about climate change. Activating fear has been an important part of campaigns to promote positive social behaviour for generations, particularly in areas like public health and safety. But can the kinds of psychological barriers I've described when it comes to fear and climate change be circumvented? Can fear work only to make people stop doing something rather than inspire broader social awareness and action?

The researchers who study fear appeals and effects are at odds about whether the benefits of using fear appeals outweigh the risks. Reviewing the literature on the effectiveness of fear appeals generally, the consensus seems to be that fear on its own won't do it. Simple fear appeals won't have lasting effects because people become desensitised to them if they're used too much—they need to be ramped up in order to overcome this and can then become too extreme, even laughable, to be credible. In addition, using fear can undermine the listener's trust in the messenger, and that can have unintended effects like denial.

In a study involving both a survey and interviews with young people living in Norwich, British academics Saffron O'Neill and Sophie Nicholson-Cole found that while fearful images of climate change can capture people's attention, they are ineffective in motivating personal engagement.[23] They left people feeling powerless, overwhelmed and fatalistic. Instead, they recommended using non-threatening imagery that connects to the everyday concerns of individuals.

Simple fear appeals may not even be effective with people who are already alarmed about climate change. In their research with Danish and Swedish climate activists, academics Jochen Kleres and Åsa Wettergren found that fear motivated action by raising awareness of the threat of climate catastrophe.[24] (As one of the activists put it, 'threat is a precondition to get up from the couch'.) But importantly, the potential downside of fear—namely, that it can paralyse people and make them despair—had to be mediated by hope that they could make a difference, particularly as part of a broad movement of people. Hope fuels action, and collective action in turn generates hope and helps reduce fear.

You can't talk about fear and climate change without addressing the elephant in the room: death, namely our own.

The work of Canadian academic Sarah Wolfe and Israeli academic Amit Tubi explores how our predictable response to our own mortality—feelings about our own inevitable death—shapes our responses to climate change. They wonder whether our slow response to environmental crises can in part be explained by this concept of 'mortality awareness':

> Like death, [climate change] is a threat that will affect everyone on the planet. Yet it is also unique because climate change is simultaneously personal and social, local and global, immediate and future, chronic and acute, known and unknowable.[25]

They found that the typical psychological defences against mortality awareness, such as denial, distraction and rationalisation, are at play when people are confronted with messages about climate change. The result could be an increase in apathy, resistance and doubt about the nature of the threat.

They also found, however, that if you made people who were already concerned about the environment more aware of their own mortality, it could actually increase their desire to act in order to leave behind a legacy for future generations. The authors describe this legacy as a 'hero project', which may include anything from throwing yourself into charity work, to seeking fame or becoming a model parent. (You might say this book is my 'hero project'.) In other words, making people aware of their own mortality when talking about climate change can make them shut down or power up, depending on how they already feel about environmental issues and their own inevitable non-existence.

From my own perspective, as someone already alarmed about climate change, thinking about it does more than make me fear my own death. Since contemplating death in my teens as a precocious young woman fond of the Cure and reading Kafka (yes, I had a black beret), I have found a way to soothe my fears by thinking less about my own non-existence and more about the legacy, if only temporary, of my brief existence on earth: my love for my children, family and friends; my professional contribution to social research and the environmental movement. I know that even though it's difficult to contemplate, my death makes way for the next generation and the generation after that. But climate change threatens all this, and with it the kind of thinking that soothes my own mortality awareness. In thinking about climate change I am thinking not just about my death but the death of everything that comes after me.

Fear is a hard emotion to sustain day after day. The fight or flight impulse is hardwired into us, priming us for immediate action but only meant to be used occasionally to save our own lives. If activated regularly, the chemicals it releases can do significant physical and psychological damage. This is why people like Tony Leiserowitz tell me 'worry' is a more productive emotion, because it doesn't hijack our cognitive abilities as much as fear does. 'Fear is not a great predictor of policy support for climate action. Worry is.'

And we know the citizens of the earth are worried about climate change. Research shows that we're now at a point where the majority of people in the majority of countries believe climate change is real and poses a real threat to future security. In 2018 a 26-nation survey conducted by Pew Research found that thirteen of these countries named climate change as the top international threat.

If only the leaders of every country would take our worries seriously.

Worry is better than fear. If we want to use fear when we communicate about climate change, we should try to combine it with positive emotions like hope, generated through collective action. Following on from this, I wonder if something like humour could work well in combination with fear when it comes to communicating about climate change, particularly with avoidant and jaded people. Like a spoonful of sugar? There are countless adages around—in everything from the Bible to *Buffy the Vampire Slayer* to *The Lion King*—about the therapeutic value of laughing in the face of danger. Maybe being funny when talking about climate could confound preconceptions about environmentalists as killjoys and wet blankets.

I first started thinking about the role of humour at the beginning of writing this book, when I had the chance to interview acclaimed scientist and Australian of the Year Professor Tim Flannery, one of our most outspoken and best known climate action advocates.

I asked him what he thought about the role of fear in climate change communication, and his response surprised me.

> Fear and anxiety are inherent in the message. The situation is scary but there are ways to mask it. There are ways we can alleviate people's anxiety while still delivering this scary message. There used to be the character of the jolly hangman in the days of capital punishment. I try to be a jolly hangman.

Talk about comic relief.

It turns out that Flannery's comment about the soothing role of humour when delivering unpalatable climate change messages is something researchers have actually looked at, especially in the United States. That research has shown there's a role for jokes, especially with certain audiences. For example, political satire (think late shows, sketch comedy, The Onion) has been shown to be effective with younger audiences who have low levels of interest in politics, making them more aware of global warming and more certain that it's happening.

Researchers at Cornell University conducted a survey of young adults, testing their responses to three videos about climate change: one that involved fear, another humour, and another that was purely informational or neutral. They found that the fear and humour appeals were equally effective in terms of promoting environmental activism.[26] Another study from the University of Colorado found that climate comedy increased awareness and engagement, as well as willingness to find new ways to solve the problems posed by a warming planet. The researchers concluded that the disarming and subversive power of comedy could help open up different thinking on an issue that can feel overwhelming and negative.[27]

Theatre maker David Finnigan doesn't need academic studies to be convinced that fear and humour can be a powerful combination when it comes to talking about climate change. Born in Canberra, the son of a CSIRO micrometeorologist, David is the author of plays entitled *Kill Climate Deniers* and the more recent *You're Safe Til 2024*. In his work, he interviews climate scientists about what's happening and might happen to the earth if CO_2 levels keep rising, then brings that to the stage with music, slide shows, monologues and some quirky dancing. 'I spend my whole practice trying to come up with funny, clever methods to talk about the things I'm scared of,' he told me.

His aim in writing these plays is not necessarily to achieve any particular effect in the audience or provoke any particular response—hope, fear, anger and so on (although that always happens). His motivation is to write about phenomena that fascinate and delight him, and as a topic climate change provides ample fodder for inspiration. He comes up with some pretty confronting and hilarious metaphors for what we're doing to the planet. For example, he has a section in *You're Safe Til 2024* about 'snarge' (you can google the derivation of that one). That's what's left of a bird after it collides in mid-air with a plane. Every day, numerous samples of this substance are scraped off the surface of planes and sent for DNA testing to help work out what kinds of birds are being 'snarged'. We don't mean to kill the birds (hey, we're just going on a work trip or on holiday) but we're killing them nonetheless. 'We aren't bad people but we kill things without meaning to,' David told me.

David's play *You're Safe Til 2024* is an iterative performance, a series of standalone pieces he writes and will perform over a six-year period, culminating in a day-long performance in 2024. From his home base, now in the United Kingdom, I asked David about his perspective on the role of fear in talking about climate. 'I'm scared obviously, but we're in it now and have been in it our whole lives. Fear is a big part of talking about climate change, but I don't live in fear.' What about the role of fear and humour? Do they play off against each other, making the fearful message more or less palatable?

> I can feel like future generations will be laughing at us and angry at us because we had the tools to fix it but we couldn't get it together. Humour in the context of talking about climate change is so beautifully inappropriate. There are legitimate feelings you're allowed to feel around climate, like anger and grief, but humour . . . well, it's like sniggering at a funeral. But that's a very human reaction. There are moments you can't help it, the ridiculousness of it all, the incredible stupidity of what we're doing right now can really hit you.

A fast-paced play about climate change that combines dark humour with even darker facts about the changing climate means David's audiences react in unpredictable ways, very much dependent on their own emotional state at the time.

> The director of *Kill Climate Deniers* told the cast before the first performance, 'Remember, with this play don't wait for a laugh because there's never going to be a predictable laugh with this show. Some people will laugh at some bits, other people at others.' And that's what happened, all of the gags had their moment, but none consistently got a laugh. It was the same with *You're Safe Til 2024*. Some people giggled at something and the others were dead silent. And afterwards people would come up and say, 'That was so depressing and bleak.' Other people, 'That was so hopeful.' Everyone has their own experience, relating to it differently, depending on where they're at.

After speaking to David, and thinking about the furious mirth of future generations he referred to, I wondered if there was a word for angry laughter. After some time googling, I came across a syndrome called the pseudobulbar affect, or PBA, sometimes referred to as emotional incontinence. It occurs when you can't help laughing uncontrollably when you should be angry (or sad). Well, laughter is one manifestation of it. The other is uncontrollable crying.

While I was putting the finishing touches on this book, WWF Australia commissioned me to do some research on Australian attitudes to the devastating summer fires. By mid-February the flames were out across the country, but the rebuilding—and reflection—had begun. There were a few surveys around showing that the fires had not only impacted large segments of the population in some way but also increased concern about climate change and the desire for governments to act.[28] At a conference I attended in Melbourne in the middle of February, a Climate Emergency Summit, many of the delegates were saying the fires had been a 'tipping point' in public attitudes about climate change.[29] Surely the extent and ferocity of the event meant that those who were disengaged, dismissive or cautious would start to take notice.

WWF commissioned me to go back to interview all the people who had been part of focus groups we'd conducted the year before looking at the issues of climate change and renewable energy. We had a pretty good understanding of how these people felt about climate change, and they tended to range from the generally concerned to the dismissive. What we found in the interviews was sobering. While a handful of people felt that the fires had made climate change feel more personal and closer, and made them more eager for better government policy, it was clear these people were already 'with the program'.

The rest, even those who believed climate change was happening, more often blamed arson and lack of back-burning. 'The fires don't even fall in the same category as climate change because it was mainly man-made out of stupidity, not due to climate change in the Atlantic or icebergs melting.' There was a palpable resentment of anyone trying to make a link between the hot weather and drought and climate change: 'I always feel they're just trying to push climate change into the news, they're using the bushfires or the drought or floods to help their cause. I don't think it's just that. There has to be other things as well. I feel like they're using people and their circumstances to push their agenda.' Instead, they tended to blame not only governments but 'environmentalists' for 'stopping back-burning' in fire-prone areas. 'I don't have any proof, but I think it's arsonists in the Rural Fire Service. Also the greenies need to back off and let them back-burn.'

All the fires seemed to do was make the concerned more concerned, the disengaged more helpless, and the dismissive and the cautious more reluctant to see the links. (They were, however, unanimous in their view that Australia should be moving faster to renewables and that governments should be driving this move.) People could disagree endlessly on why these fires had happened at the same time as they could furiously agree on the main solutions to reducing emissions.

We can't hope for a tipping point, for the bad weather to deliver a 'light-bulb' moment for the whole population. The cognitive barriers won't be washed away by floodwaters or burned away by bushfires. But the solutions, well, everyone can agree on those, even when we don't completely understand the science behind a solar panel.

There are those who argue that fear is a spent force when it comes to convincing the cautious and undecided about the urgency of climate action. They contend that many of us have 'apocalypse fatigue', that we're so overly familiar with fictional stories about the end of the world on TV, movies and in novels that the notion they might all offer accurate predictions of the near future seems as far-fetched as Daenerys Targaryen swooping down on one of her dragons to carry away the family dog.

I have seen this incredulity about the apocalyptic proportions of climate change play out in my own research, with dismissive and cautious participants reflecting on all the times we were supposed to fear the end of the world via, say, nuclear war or the Y2K bug. (At press time, the jury was still out on COVID-19.) But if the research to date is any fair measure, fear has its part to play, if only in combination with other emotional appeals to hope, humour and collective action.

There is one group of people I want to see absolutely terrified all the time. That's the politicians across the globe who have dragged their feet on climate action. I don't want them just to fear climate change. I want them to fear the wrath of the citizens who trusted them to serve and protect.

I want them to fear us.

CHAPTER 6

ANGER

Or how to turn anger
into activism

When Anna Oposa gave me her business card, it had 'Chief Mermaid' as her job description. But she didn't need a cute moniker to be memorable. I met the glamorous ocean conservationist when she had flown (not swum) from her home in the Philippines to speak at the Climate Reality conference.

One story she told at that event stood out for me. She had been leading a campaign called Defying Gravity, the aim of which was to stop a popular but harmful practice in her country of tying prayers to helium balloons and releasing them into the sky. At one annual festival in her province, hundreds of these prayer balloons would be released. They would eventually fall into the sea and add to the already astounding amount of plastic floating around our oceans, endangering sea life.

But she was having difficulties convincing people that the lanterns were a problem. 'I did one interview with a journalist who said to me, "Who are you to tell people what to do and believe? Don't you know when you release the balloons, the prayers reach God?" I told him about, you know, gravity, and I think all the balloons fall down. But he kept telling me, "You shouldn't be questioning other people's beliefs."' Instead of arguing further with the journo, Anna took a different approach. 'I don't believe in making fun of people's beliefs. I am always thinking, "How can we accept them as a given and get them to change anyway?"'

So the NGO she helped found, Save Philippine Seas, wrote to Catholic Church leaders (her country has the third-largest Catholic population in the world) to ask them if they would speak out about the practice. The letter emphasised that we are stewards of the earth as God's creation and should protect its habitat and creatures. 'They never replied but they did send out a press release very soon after and said they were banning the practice, repeating a lot of what we said to them.'

While Anna and her organisation can't take credit for changing the church leaders' minds, the timing was there. It was a win that confirmed for her that despite differences in opinion about religion and science, 'there are ways to campaign to find mutual points of empathy'.

Anna grew up in a house where climate change was an ordinary dinner-table topic. Her father is an international environmental lawyer, and when she was at university he got her involved as a researcher on a global legal action on climate change. At the time, stronger storms and bigger floods were hitting the Philippines, causing devastation. As I mentioned in an earlier chapter, when she attended a United Nations climate change conference as a youth delegate, she was astounded to learn that denialism was so strong in countries like Australia. She wrote an advocacy handbook for young people about climate change, funded by the Philippines Department of Education. 'There were no local materials around, only materials from Western countries with lots of pictures of polar bears. That's not the symbol of climate change in my country. It's people losing their houses to floods or wading through waist-high water.'

As Anna and I skyped one morning (she in her study in Manila, I in mine in Sydney), I asked her how she managed her emotional reaction to climate change. In particular, how she coped with feelings of frustration and anger. 'As a young advocate, I used to cry a lot,' she told me. 'I would cry about every single issue. "Why aren't people listening to me?" And then I realised, if I wanted to stay in this advocacy space, I needed to learn how to manage my emotions.' But it was advice from her father, who'd been fighting the good fight for many years, that helped Anna come to better terms with her anger.

> There was one time I had a really bad experience with a government leader and I called my dad in tears. And he started laughing at me. He said, 'How long are you going to cry? You can be sad and angry for a few seconds and then you have to realise, what are you going to do next? You can't think of your advocacy as a battle, because if you do, you're always going to feel like you're going to war. You're always going to be angry, full of all these negative

emotions. As if it's always win or lose. Instead of a battle, you should think about it as a game. If you're thinking of it as a game, you get creative. You learn how to plan for the long run and you might actually enjoy it. If you think of it as a battle, you're always going to be the angry one, the combative one. No one wants to be around the angry one.'

Anna took her father's sage advice to heart and started to approach her advocacy work with less anger and more good humour. She tells me that despite climate change and all the other problems facing her country, people in the Philippines are happy people. And so a lot of her advocacy work involves creating funny and quirky memes promoting environmentally sound behaviour, which she argues is both effective at getting the message across and helps makes her advocacy enjoyable. 'Anger is still important. There are days I get very upset. And yes, there's resentment that my country is suffering because other richer countries won't do more. But that's never a reason not to act.'

Anger gets a bad rap. It's almost always seen in negative terms. Many, many cultures view anger with suspicion, as a primitive, unsophisticated, antisocial and dangerous emotion. Anger comes when we encounter some kind of offence or when we recognise that one of our goals is being frustrated by someone or something.

Like fear, the main goal of anger is to motivate us to tackle a threat. Indeed, anger and fear can be prompted at the same time in response to the same threat, like a terrorist attack or a home invasion.[1] Anger is also related to other negative emotions like guilt. Most of all, we see anger as undermining our more constructive decision-making abilities. As Leonie Huddy and her colleagues from Stony Brook University point out, anger is often associated with lower levels of cognitive effort. Anger can encourage certain behaviours in people, such as being less careful and systematic in their thinking—so much so that they can be less aware or concerned about the consequences of risky actions.[2] Angry people are almost always seen as irrational. They punch strangers in bars, run people off the road with their cars and say things they don't mean. Groups of angry people are even more scarily unpredictable. They can burn their own communities to the ground, storm buildings and string people up from the nearest tree.

But like all human emotions, anger serves its purpose in certain circumstances. As is the case with fear, we wouldn't be able to protect ourselves, the people and things we love, if we couldn't feel some degree of anger. Daniel A. Chapman and his colleagues from the University of Pennsylvania argue that while anger is often seen as a destructive emotion leading to aggression, it in fact rarely leads to actual violence. While links between anger and violence obviously exist, they operate in complex ways depending on the context in which they occur. Instead of an overly simplistic view that anger is destructive, they cite research showing that anger is typically the emotion most associated with inspiring people to address social justice issues—discrimination, for example.[3]

In this way anger, channelled effectively, becomes a positive and productive emotion, especially when it comes to turning concerned citizens into activists. The galvanising capacity of anger is well respected and understood by people who study social movements throughout history. The positive power of anger is summed up beautifully in this quote from John Bird, life peer of the House of Lords and founder of *The Big Issue* magazine.

> Anger is a brilliant initiator. If you don't feel angry about the world you probably don't want to change it . . . As a child, I always remember sitting by the gas oven and my mum would put water in and porridge in and tell me to stir it. I would think that stirring would last for ever, but at some point it would turn into porridge you could eat. It's how I feel with anger—you've got to make it something thicker. You want it to be fuel for a rocket, not a gun.[4]

Of course, while anger can shape and inspire activists, it can also lead to burnout among leaders, and internal conflicts within organisations and social movements (as anyone who has ever been involved in these groups can confirm). And so researchers have shown that anger needs to be coupled with other emotions to be that effective rocket fuel Lord Bird describes.

For example, in their research on Danish and Swedish climate activists, Jochen Kleres and Åsa Wettergren found that anger was a critical yet complex emotion for the young men and women involved in climate-related NGOs.[5] Anger was a powerful political emotion that allowed protesters to focus their attention on those in power responsible for climate damage. That being said, the young participants recognised that presenting an angry face to the public all the time was not the way to mobilise support for their cause (as Anna said, no one wants to be around the angry one). And so they tried to avoid expressions of anger in their advocacy.

Ultimately Kleres and Wettergren concluded that anger, hope and fear were all important emotions for getting activists involved and then, more importantly, keeping them involved. 'Anger makes hope "a strong energy".' And 'hope makes anger "positive".' Arguably, a fruitful combination of anger and hope may defuse the destructiveness of anger.

But what about the rest of us? If you're not already a member of Extinction Rebellion, are angry messages about climate change going to work?

When I think about anger and climate change, my thoughts immediately turn to Twitter. An increasing number of the people I follow on Twitter are involved in climate politics or are interested in climate change. Scrolling through my feed every day, amid the retweeting of useful news articles, I watch a seemingly endless stream of abusive 'back and forth' about climate causes and effects. It's predictable, painful and often boring, even for someone like me whose job it is to keep across this stuff.

I've always suspected that the Twitter wars on climate change serve to further entrench the politicisation of the issue, turning less committed people off the topic and discouraging them from engaging. This is certainly George Marshall's view. He argues that social media has become the place where people aggressively express their in-group identity on political issues, with climate change a prime example. By 'in-group identity', he means not only identifying strongly with the social group we belong to but believing that group membership makes us superior to other groups. He argues that in reviewing the swarms of comments about climate across social media platforms, you can see this 'in-group performance' at work.[6]

The aim of getting into fights and trading insults about climate change on the internet is not to change anyone's minds, Marshall argues, but to reinforce already existing views and show your tribe that you're a loyal keyboard warrior. Ultimately, he questions whether the emotional energy required to get into fights on social media about climate is worth the pain given 'the vast majority of people are being entirely ignored during [these] punch-ups'.[7]

There has been some interesting research into the impact of social media on people's attitudes to climate, and the results, unsurprisingly, are mixed. Emily Cody and her colleagues from the University of Vermont analysed Twitter mentions of climate between September 2008 and July 2014. They concluded that most of the Twitter traffic about climate change is actually from green activists rather than deniers, which makes it an important resource for spreading news and general awareness.[8] Ashley Anderson, an academic at Colorado State University, was less positive, however, in her 2017 study.[9] She found that the tone of climate change coverage on social media was mostly negative, with the potential to turn audiences off. 'While there is reason to be optimistic about the ability of social media to positively influence opinion, knowledge, and behaviour around climate change . . . social media use may simply encourage reinforcement of existing perceptions of climate change rather than reaching new individuals or changing opinions.'

The impact of social media 'discussions'—whether genuine exchanges between real individuals or bots, or manipulation by professional and amateur trolls—is such an under-researched and fast-moving area it's almost impossible to draw any firm conclusions at the moment. What's more, my experience is that climate change conversations play out differently on different platforms—Facebook as opposed to Instagram or Reddit—with people heavily curating the information they might want to receive and the people they might want to interact with (my experience of climate change information on Facebook must be vastly different from that of someone whose major interests are, for example, surfing, cars and construction).

I've found out some of the most interesting information about climate change via social media, mainly through the posts of scientists and writers from all around the world. But I can also say that in the research I do with Australians, the spats they observe on social media over climate have two main impacts: turning people off the discussion, and reinforcing the view that common ground on climate is hard to find.

This leads us nicely to the main downside of anger as an emotion, namely its tendency to create an 'us' and a 'them'. Heroes and villains. Enemies and allies. As I mentioned before, anger is a response to an offence caused or to the frustration of a goal. Anger is increased when the offence or frustration is regarded as somehow unfair or unjust. And in such cases there's usually a person, organisation or system unfairly offending or unjustly frustrating. Given that anger is often, in the heat of the moment, directed at someone or something, it can easily hit the wrong target (think of how common it is to take your anger out on an inanimate object—it's not the ironing board's fault your partner constantly leaves dirty clothes on the floor). An angry person can easily find a convenient cause or scapegoat to blame for their predicament.

A number of studies have shown that anger can increase our impulse to condemn and punish others, and even diminish our own sense of responsibility or need to act. In their survey of more than 700 American adults, academics from Cornell University, Hang Lu and Jonathon P. Schuldt, wanted to test the idea that anger increases our tendency to blame others for social ills. They found that stimulating anger against fossil fuel industries in participants increased their support for government policies to make these industries more responsible for their carbon omissions. 'When some participants were primed to feel angry, they more strongly endorsed behavioural modification intentions that entailed punishing others,' such as consumer boycotts.[10]

The capacity for anger to increase our desire to blame and punish was confirmed in another study by Leonie Huddy and her colleagues from Stony Brook University. Looking at the impact of anger and anxiety on American voters' support for the Iraq War, they found that anger led participants to minimise the risks of a war and increased support for military action (as opposed to anxiety, which heightened perceived risk and reduced support for war).[11]

The blame game is an almost inevitable flow-on effect of our angry response to unfairness. During Australia's 2019–20 Black Summer, when fires were still burning around the country and communities contemplating how to rebuild, the media was full of finger-pointing. While environmentalists were blaming politicians and the media, climate deniers in our media and even some government figures were blaming environmentalists and arsonists. Ordinary citizens pointed the finger in all directions. It was hard to know what precise consequences would flow from all this anger and resentment, especially when it came to politics, given state and federal elections were some way off. Would public support swing towards more aggressive climate change policies or the opposite?

While anger can forge activists and progressive social movements like the environmental movement, it can also see the election of populist governments, which globally have been characterised by recalcitrance to act on climate change, if not outright denial. Indeed, populist politicians have shown themselves to be expert at funnelling people's anger against scapegoats (people of colour, immigrants and refugees in particular). They are masters of the blame game. Stoknes argues that the impulse to create this division between the 'righteous us' and the 'evil them' is widespread in Western culture after 2000 years of Christianity and monotheism. This assumption of moral opposition is mostly destructive, further entrenching polarisation and fundamentalism, damaging the cohesion and cooperation necessary to tackle the problems we face.[12] George Marshall agrees, arguing that the hero/villain approach gets in the way of emphasising cooperation, mutual interests and our common humanity.[13]

More than that, as we saw in the chapter on guilt, attributing blame for climate change is not an easy task. Certainly there are those who rank higher on the blame scale than others (with great power comes great responsibility). But many of us in affluent countries have made our own contribution, either through our own behaviour or by voting for governments with a shocking disregard for climate change and its impact on other nations.

And this is where my attitudes about the role of anger in the climate change discussion become more complicated. Who am I to tell the people of the Torres Strait, or Fiji, or Bangladesh, or the Philippines to tone down their anger because it isn't productive? That it's getting in the way of finding common ground and mutual interest? From where they stand, it's sink or swim, and it doesn't look like my country and its leaders care much about them at all. At the very least I can let them feel angry about that.

And while we might like to tell ourselves that people in the Pacific are happy-go-lucky, when it comes to climate change, believe me, they are also bloody angry. In a comparative study of residents of four island nations—Fiji, Cyprus, New Zealand and England—American academic Margaret V. du Bray and her colleagues found that the residents of Fiji (facing rising seas) and Cyprus (facing crippling water shortages) were most likely to express anger over climate change.

> Climatically disadvantaged groups are disproportionately likely to experience emotional distress as a result of environmental injustice. Of all the groups, the Fijians—most immediately physically and economically vulnerable to the effects of climate change, and most emotionally connected to the specific place in which they currently live—are the most emotionally distressed.[14]

Think about Lavetanalagi Seru's poem I quoted in Chapter 4: '*Loss of land . . . / Caused by . . . humans / at whose hands, also lie the cure*'. Creating a 'righteous us' and an 'evil them' can be problematic when you're trying to argue for climate change action. But that doesn't mean Lavetanalagi's anger isn't righteous.

Even writers like George Marshall and Per Espen Stoknes recognise that subduing our feelings of anger is hard. Those feelings do need to be let loose from time to time. 'Maybe my anger needs to cry without being impatiently and prematurely pushed and bullied into positive thinking, quick fixes and social movements,' Stoknes writes.[15] Indeed, for those of us already engaged with the climate issue and for people becoming more interested, shutting out these feelings of anger, frustration and rage is both futile and naive. They are now an integral part of the new emotional regime of the climate age, and we have to find a way to live with them.

In her work at RMIT in Melbourne, Australia, academic Blanche Verlie found that during their twelve-week uni course on climate, students went through a range of emotions. They felt at different points angry, resentful, challenged, confronted and dismayed. Verlie describes these emotions as part and parcel of 'the process of learning to live-with climate change'.[16] It's something all of us, activists or not, need to learn how to do quickly, if we're not already.

In his extraordinary book *The Righteous Mind: Why good people are divided by politics and religion,* psychologist Jonathan Haidt reflects on the benefits of letting go of anger, particularly anger generated by political partisanism. Haidt was an ardent liberal when he was a young man studying at Yale, railing against Reaganism and 1980s neoliberalism in his classes and among his friends. Later on in his career, he left America, travelled through countries like India, and studied cultural psychology. He then returned home and felt able to listen to and understand the arguments of conservatives without the kind of knee-jerk, hostile reactions he had felt as a young man. Instead of dismissing conservative arguments out of hand, he was able to view them with more circumspection 'as manifestations of deeply conflicting but equally heartfelt visions of the good society'.[17] He didn't cease being liberal in his views, but he became more open-hearted in his reactions to other people's beliefs, an important distinction.

'Once I was no longer angry, I was no longer committed to reaching the conclusion that righteous anger demands: we are right, they are wrong.'[18]

We saw in a previous chapter how the political polarisation of the climate change issue has become entrenched over the last few decades as climate change has become a more serious threat and its manifestations more obvious. Certainly, breaking down partisan divides, this us versus them mentality, is important when trying to convince people that the vast majority of us in every country in the world will be better off dealing with the climate threat. But again, that doesn't mean there isn't a hierarchy of blame for the causes of climate change. Just because I use a hairdryer and take flights for work doesn't let fossil fuel magnates like the Koch brothers off the hook.

As I said at the start of the chapter, anger gets a bad rap, and unjustifiably so (and if anger was a person, like in that Disney movie *Inside Out*, he would be pretty angry about that). Like fear, anger about climate can feel justified, appropriate and even at times cathartic. That's why I like to look at anger as a kind of 'frenemy'. Too much time spent together isn't good for you, but ultimately the connection keeps you on your toes and makes you strive to do better. Instinctively, I've always thought that constructive flashes of anger—personal and collective—are essential in any drive towards addressing injustice. Rocket fuel that can help topple governments.

We can, however, be angry and frustrated from time and time without seeing climate change as a war between us and them.

Marshall argues that instead of looking at those trying to derail or undermine the climate cause as enemies, we should see them as 'obstacles' (an important albeit delicate distinction, I admit).[19] And, as Anna's dad suggested, instead of seeing our efforts to deal with climate change as a war we should see them a game.

A really, really long game.

The last three chapters have explored the companion emotions of guilt, fear and anger. I hope I've shown you that even though these are often characterised as negative emotions and can have negative effects when it comes to talking about climate change, they are necessary and even helpful to the climate cause. And while there are no hard and fast rules when it comes to talking about climate change—one part anger and three parts hope, with a sprinkle of guilt, for example—in general, positive emotions like pride, empathy and compassion are more effective, especially with people who aren't already 100 per cent convinced about the need for action.

We now head down a darker path in the next two chapters, to explore the outer reaches of our emotions on climate: denial and despair.

CHAPTER 7

DENIAL
Or the need to be innocent

I experience something disarming when people in the focus groups I conduct for a living deny that climate change is real, dangerous or being caused by humans. I *want* to believe them. It's something that doesn't happen when listening to views I don't agree with on any other topic. But when people who don't believe in climate change say that the future will be much the same as the present or that the threat is exaggerated, I feel drawn towards this comforting vision. It's like a warm bath I want to soak in to forget the worries of the world.

In the social science studies into climate change and emotion, denial receives a lot of attention. The allure of the topic for people like me is no surprise. What's up with deniers? What's wrong with them? Are they stupid? But also, speaking personally at least, there's a tiny bit of envy. Wouldn't it be nice not to believe in climate change? And every now and then, after I put down a particularly upsetting article about climate effects or scenarios about what will happen when we reach 3 degrees

Celsius, for a moment I say to myself, 'Perhaps, maybe, it won't be that bad.'

In psychology, denial is when you refuse to acknowledge the significance or consequences of your behaviours or refuse to accept the reality of a situation. We often associate it with people addicted to alcohol or drugs who continue to maintain they haven't got a problem (even though they're clearly harming their health, family and career). But denial is a term unfit for purpose when it comes to describing the people who argue that climate change isn't happening or is benign or is a purely natural process. That's because denial has many faces. And not all kinds of denial are equal.

Climate denial is a term used widely and applied broadly, almost always as an insult (which is probably why many powerful individuals who do deny the reality of climate change prefer to call themselves 'sceptics'). The 'denier' label hides a number of important distinctions that need to be drawn out. The first one Per Espen Stoknes points out is the difference between active denial and passive denial.

Active denial exists when someone energetically engages with the issue, eager to debate, refuse and refute the existence of climate change with friends, co-workers and strangers on social media.[1] Passive denial suggests the opposite, of course: that when faced with climate change as a topic, the person responds with indifference, saying they don't care or it's not important. Or they just shut down. As Stoknes writes, for those people engaged in passive denial, 'the issue becomes unspeakable [and denial] offers itself as a convenient way out of the discomfort'.[2]

In the focus groups I conduct, I encounter more passive than active denial. Active deniers, well, you can't get them to stop talking. Passive deniers rarely say more than a few words at a time, rarely unprompted, and seem particularly interested in eating the sandwiches we provide as part of the focus group experience.

The next critical distinction is between true denial and scepticism. As I just mentioned, a lot of climate deniers prefer to describe themselves as sceptics. It sounds more reasonable, like you are someone who has *really* looked into this climate change business and still has some important and legitimate questions left unanswered. But here's the thing. A lot of people who describe themselves as sceptics are not in fact sceptical at all. In order to understand why, we need to return to a topic from earlier in the book: the nature of the scientific method.

Part of the scientific method is that any form of knowledge has to be tested and retested to be accepted as scientific truth. And something can be part of the accepted body of knowledge in science and still be open to questioning, perhaps not at the level of fundamental principles but in terms of modification and extension of established understanding. In this way, as Stoknes puts it, 'science itself is really systemized skepticism', the constant questioning and testing of scientific knowledge conducted under agreed and rigorous conditions within the academy. Being a sceptic is, in science, a good thing.[3]

And there have been true climate sceptics who have questioned and even rejected the idea that humans are causing the planet to warm quickly with devastating consequences for us all. Many of them in science, politics, industry, government and in disciplines like meteorology have now changed their tune given so much of the early climate change modelling has proven to be accurate.

But the people who argue climate change isn't happening or isn't being caused by humans aren't engaged in scientific scepticism or even constructive questioning. They're engaged in deliberate, even belligerent, cherry-picking of the information they want to hear. And, as you will see later in this chapter, it has nothing to do with the search for scientific truth.

So we have active and passive deniers and true sceptics (albeit their numbers are diminishing), and deniers masquerading as sceptics.

The final and important distinction is between what I would call professional and amateur deniers. Professional deniers are in positions of power in societies around the world—politics, industry, the media, and influential and well-resourced think tanks. They make their living and derive their status from attacking the climate change science, which can include attacking climate scientists themselves.

Amateur deniers, well, they might vote for the denying politicians, consume denial media and follow people on Twitter who promote denying messages, but they don't earn a living from climate denial. This is the difference, say, between Rush Limbaugh and your drunk uncle you don't like to sit next to at family gatherings.

Professor Andrew Hoffman from the University of Michigan makes a similar distinction in terms of the institutional support for denial in America, which applies just as well to countries like Australia. He is careful to distinguish between the organised 'climate denier' movement and the broader 'sceptical' population:

> Whereas the organized denier movement is a collective social movement run by professional advocacy organizations working to discredit climate change like the Heartland Institute and conservative think tanks like the Cato Institute that produce research and white papers, the 'skeptical' label is ascribed to a population who are doubtful about climate change or the motivations behind calls for climate action in the broader population.[4]

Why does it matter what kind of denial someone is in or what kind of denier someone is? Well, it matters because the more active and professional a denier is, the more power they have to slow down any attempts to deal with climate change. And they have proven to be extremely successful in these attempts.

Given we have a sense of the spectrum of denial out there, what's driving it? Fear is a good starting point. The social science research indicates that climate change denial is a coping mechanism to deal with the fears triggered by knowledge about climate change. But that seems like an inadequate explanation.

As we saw in the chapter on fear, climate activists often fearfully respond to climate change information and have to learn to push those fears aside in order to function. Fear in response to climate change is a universal and understandable reaction. Something more than just fear must be at work when we deny climate change. As Stoknes writes, 'when we negate, ignore, or otherwise avoid acknowledging the unsettling facts about climate change, we find refuge from fear and guilt'.[5] This is the case with passive denial in particular, and we all engage in this from time to time.

But something else is happening when we engage in outspoken denial and mockery of climate science—in other words active denial. Active denial allows us to defend ourselves against those we feel are criticising our lifestyle, who believe they have the right to tell us how to live. Here we can see denial not as a lack of information or intelligence but 'self-protection'.

This is the most important aspect of denial to grasp when struggling to understand the phenomenon. People who deny climate science are not stupid or unable to find the right information. They are reacting to a message that threatens their worldview, their values and even their sense of self. Stoknes describes the roots of denial elegantly, saying we deny climate change because of our intense need to be innocent. 'Deniers feel a need to defend their identity and lifestyle against the message that climate disruption is real, urgent and caused by human fossil fuel use; they feel an inner need to explain it away.'[6]

As we saw in the chapter on guilt, implicit in the climate change message, especially for wealthy people in wealthy countries, is that we are causing and benefiting from a process that will damage the world and harm billions of people. We don't want climate change to be our fault (we want to be innocent) and so we deny that it's happening or happening to the extent it is or that there is anything we can do about it.

If we think of it in this way, debating the science in order to change the mind of a climate denier seems destined to fail, because when you're trying to convince them about the facts of climate change you're actually just, as Stoknes puts it, 'crashing against the wall of the self'.[7] You're asking them to change their view not just on climate but on everything that matters to them. You think you're arguing about the laws of physics and weather patterns, whereas you're really debating competing visions of the past, present and future, and of the nature of authority and power in our society. And because climate deniers start with a conclusion and single-mindedly work backwards (which is something we're all guilty of at times), their commitment to coherence in their position is minimal at best.

In their analysis of the arguments from climate deniers, researchers led by academic Stephan Lewandowsky from the University of Bristol found that a rejection of climate science involves arguments that are piecemeal and disjointed rather than systematic.[8] In the world of climate denialism, statements like 'the globe is cooling' can coexist with claims that 'the observed warming is natural' and that 'human influence does not matter because warming is good for us'.

If we understand climate denial not as rejection of the science but as a response to a threat to identity, status and power, then it makes sense that, as so many studies around the world have found, certain demographic trends are discernible in beliefs about climate change. In other words, certain kinds of people are more likely to deny climate change than others because their identity is more threatened by the climate message.

If acceptance of the reality of human-induced climate change opens up questions of identity, power and privilege, well, those at the top of the pyramid are predictably likely to feel more than just a tad resistant to accepting the reality of climate change. In the United States, research by Aaron McCright and Riley Dunlap has shown that conservative white males are more likely than other groups to hold sceptical or denialist positions.[9]

There is also a growing body of research showing strong links between climate denialism and anti-feminism. Researchers at Sweden's Chalmers University of Technology, which recently launched the world's first academic research centre to study climate denialism, have for years been examining the links between climate deniers and the anti-feminist far right. In a 2014 paper looking at the language of climate sceptics, academics Jonas Anshelm and Martin Hultman found that there were common themes in anti-feminism and climate denialism.[10] Caring about the environment was seen as weak and womanish. Mastering the environment, accepting the possible risk of climate damage, was associated with masculine strength.

While there are other common demographic skews in the characteristics of those who deny climate science—older rather than younger, men more than women, conservative more than progressive, country more than city—we can't let these general trends lead us to make too many generalisations. Or worse, indulge in stereotyping. One of the more damaging stereotypes about deniers is that the less educated you are, the more you doubt climate change.

To be fair, there is some evidence out there to support this. Many of the surveys I've conducted in Australia on attitudes to climate as well as research conducted by others show that those who reject the climate science are likely to be less educated than those who are highly concerned about the issue. This education attitude gap on climate exists in other countries as well.

But this should not be overstated, because it plays into our perceptions that belief in climate change is all about understanding the science (whereas in fact we know it's more about worldview than intelligence). Low levels of formal education are not an impediment to belief about climate change in places like Fiji or the Philippines. And indeed, some American research has shown that with the climate issue becoming more politicised, it is actually *educated* white conservatives who are more likely to deny or resist climate messages rather than people with lower levels of education.[11]

We know that different kinds of fear drive different forms of denial, but what else do we know about climate denial in societies where it thrives? First of all, we know that people who believe humans are causing the climate to change and are concerned about that outnumber the people who don't.

A 2019 study by market research company YouGov found that people who deny climate change amount to 10 per cent of the population or less in countries like Mexico, Australia, Germany, China, France and Britain. Indonesia (18 per cent), Saudi Arabia (16 per cent) and America (13 per cent) had higher percentages of deniers in their populations.[12] In his 2018 study of deniers in thirteen countries, Australian academic Bruce Tranter found his own country topped the list when it came to climate denial, at 22 per cent, equal with Norway, followed closely by the United States, New Zealand and Finland.[13] I have seen countless studies of attitudes to climate change in Australia that put the percentage of climate deniers anywhere from single digits to 22 per cent.

Regardless of what definition of 'denial' you use and which countries you include in your study, rejection of the climate science remains at around less than a quarter of the population. But, as Tranter points out, this could still be described as 'a substantial minority'. Substantial in that denialism can still translate into political power, depending on the structure of a nation's political system. For example, people who deny climate change can, with a very small percentage of the popular vote, find themselves elected to upper houses in the parliamentary system in my home country and go on to block climate legislation or work with conservative parties to wind back environmental reform. Or they can cluster as a faction in larger political parties and again slow the pace of reform on climate.

And more important than the percentage of deniers in the community is their influence on public perception of the climate issue. A number of studies around the world have shown that people tend to overestimate how many people reject the existence of climate change. For example, between 2010 and 2014, Australia's government research agency, the CSIRO, commissioned a project involving nearly 17,500 Australians. It found that just under 80 per cent of respondents thought climate change was happening and, on average, 62 per cent of those people believed human activity accounted for these changes.[14] But when it came to predicting the views of their fellow Australians, the participants *overestimated* the ranks of deniers in the community. On average, they predicted that 23 per cent of Australians were of the opinion that climate change was not happening, when the actual number in the survey was 8 per cent.

Not only is there a tendency to overestimate the number of climate deniers in the community, but we seem to overestimate the numbers of climate deniers in the scientific community as well. Research in the United States found a whopping 87 per cent of Americans were unaware there is a scientific consensus on climate change.[15] So while we might look at the actual number of deniers in the community as measured by polling and feel some sense of relief, we have to consider the important amplification effect of professional denialism on our perceptions. It makes us feel like there is more uncertainty—in both the science and the community—than there actually is. (It doesn't help that denialism is aided and abetted by parts of the media that keep putting deniers on platforms with climate scientists, supposedly in the interests of balance.)

Indeed, if deniers have a rhetorical superpower, it's their ability to exploit human discomfort with uncertainty. Again, this leads us back to the challenges the scientific method poses for lay people looking to understand climate change. As British academic Mike Hulme points out,

> uncertainty pervades scientific predictions about the future performance of global and regional climates . . . Some uncertainty originates from an incomplete understanding of how the physical climate system works . . . Other sources of uncertainty emerge from the innate unpredictability of large, complex and chaotic systems such as the global atmosphere and ocean . . . A third category of uncertainty originates as a consequence of humans being part of the future being predicted . . . The best that can be done is to work with a range of broad-scale scenarios, and range of possible futures.[16]

Professional deniers take this scientific uncertainty and sow the seeds of doubt in people. The scientists are uncertain, they say. The science is still being debated. There is no community agreement. We need more time. No need to act right now.

Given everything we know about denialism and deniers—that they're more powerful than their numbers in the community and that arguing with them is like trying to shove an octopus into a string bag—what, then, do we do? Is it even worth arguing with a denier?

Many researchers and thinkers who have delved into the minds of deniers say don't bother. Stoknes argues, 'it's no use trying to win the argument against those that have made up their minds to the contrary'.[17] Marshall shows how professional and skilled denialists (think Marc Morano in the United States or Lord Christopher Monckton from the United Kingdom) aren't interested in debating climate change through an exchange of ideas but see it more like a 'debating slam', the verbal equivalent of cage fighting.[18] I myself feel torn about the issue. I know I can't change the mind of a climate denier, but given the impact the professional deniers have on the views of others, I don't like the idea of ignoring them either.

A friend and colleague of mine who has valiantly—and very publicly—tried to change the mind of a climate sceptic is Anna Rose, one of Australia's best known climate activists. In 2006, at the age of 23, she co-founded the Australian Youth Climate Coalition, a movement of more than 70,000 young people. She has worked for, written about and fought tirelessly for the cause of climate change since her teens. She grew up in one of Australia's famous coal ports, Newcastle, and her family has run wheat farms in the north-western regions of New South Wales for generations.

Her schoolyard activism around climate change was sparked by a combination of the school curriculum and her first-hand experience of drought on her family farm. Softly spoken and almost shy when you meet her in person, once she takes the stage she is a passionate and persuasive climate communicator.

She is also a climate nerd, across the latest climate science as well as all the policy implications of acting on climate change. Exactly the kind of person you'd choose to debate a climate sceptic on television.

And in 2011 that's exactly what happened. Anna was chosen by the ABC, Australia's national broadcaster, to feature in a documentary called *I Can Change Your Mind About Climate Change*. The one-hour show involved Anna travelling around the world with a TV crew, accompanied by a recently retired conservative politician, Nick Minchin. Talk about the odd couple. For many decades Minchin had been a federal senator, government minister and powerful leader of the conservative party's 'dry' or right-wing faction. He was an outspoken 'sceptic' about climate science and, when he was in government, determined to argue at every turn against effective policy on climate change.

The premise of the documentary was that Nick and Anna would each pick a group of experts and visit them together, in the hope that these experts would be able to convince the other of the truth (or not) of human-caused climate change. The documentary makers themselves would also throw some apparently 'neutral' and surprise guests into the mix. The world tour would take them from the top of a Hawaiian volcano, to the streets of Washington, DC and London, the suburbs of Perth and the eroding cliffs on the eastern coastline of Britain.

Anna thought long and hard about getting involved in the documentary. Other environmentalists had been sounded out and refused. They cautioned her against involvement. 'They thought

I was naive to think I could "win" the argument; that the whole idea of the show played into the denialists' strategy of framing the science as disputed when it actually wasn't.'[19] But she decided to go ahead. 'This was a chance to reach a large number of people—directly in their living rooms—who the climate movement have struggled to communicate with for years.'

Anna approached her part in this roadshow with characteristic hard work and good faith. She spent every available hour on planes and in hotel rooms inhaling as much of the science as she could. She picked her experts with some clear criteria in mind, namely respected scientists and experts rather than argumentative, left-wing activists. For example, she chose Admiral David Titley, chief oceanographer of the US Navy working out of the Pentagon, a former climate sceptic who now accepts the science and leads the navy's task force on climate change. 'On any other matter, Nick would take his views extremely seriously,' she wrote.[20]

And she also picked conservative British politician and Brexiteer Zac Goldsmith, who makes a compelling case for serious action on climate as a huge economic opportunity rather than merely an excuse for big government. 'I chose my spokespeople because they could reach out to Nick and have a chance of actually changing his mind,' Anna explained. 'They're moderate and polite. If anything, they understate the seriousness of climate change in order not to scare off conservatives like Nick.'[21]

Nick Minchin took a different approach. He chose 'experts' like Joanne Codling, who had authored booklets entitled

The Skeptics Handbook and *Global Bullies Want Your Money*, and her partner David Evans. Neither has any formal qualifications in climate science. The controversial climate denier Marc Morano, who has stated climate scientists should be 'publicly flogged' and that climate change is a communist hustle. Professor Richard Lindzen, who continues to argue that clouds will suppress global warming (and that the science behind the dangers of second-hand cigarette smoke is bogus). And the late Christopher Booker, a journalist who disputed evolution and believed white asbestos is safe.

Between visits to these people, Nick gave Anna right-wing magazines like *Quadrant* to read, with articles arguing that environmentalists hate human beings and are communists who want to restrict freedom and punish people for the way they live their lives. He also included Bjørn Lomborg, who has shifted during his career from being an ardent climate denier to a 'lukewarmer' (someone who admits global warming is happening but slowly) to someone who says he agrees with the science but argues against measures like a carbon tax.

Minchin also suggested they visit the Geneva headquarters of CERN, the European Organization for Nuclear Research, to hear about early research on the role that cosmic rays from outer space might play in global warming. (But even the lead researcher rejected the way his project was being interpreted by climate sceptics as evidence that something other than CO_2 might be causing climate change, dismissing it as an extreme point of view.)[22]

Minchin might describe himself as a sceptic, but his choice of experts was straight from the guidebook for science denial. In their influential work on the characteristics of science denial, Mark and Chris Hoofnagle argue that popular tactics of denialists include conspiracy theories ('climate change is a hustle'), fake experts (individuals purporting to be experts but whose views are wildly inconsistent with established knowledge) and cherry-picking (selectively drawing on isolated studies and not looking at the broader body of research on a topic).[23]

In choosing her diplomatic experts with PhDs in science, Anna was actually attempting to change Nick's mind on the basis of the evidence. Nick's aim was mostly to find charismatic contrarians who would further sow the seeds of doubt in the public's mind. He didn't seem at all fussed about whether Anna listened to them or not.

Anna and Nick's long conversations on their road trip revealed the real driver behind the former minister's climate scepticism, which reflects all the social science research outlined in this chapter. Nick is a clever man who understands what the climate scientists are telling him. But it's the idea that humans have to radically change their behaviour in response to that science that's the real obstacle. 'Perhaps he accepts science that doesn't challenge his world view but rejects science that does,' ponders Anna about halfway through her journey.[24] Minchin's commitment to minimising government regulation and maximising the power of the free market is non-negotiable. Therefore the science, however well supported by data and advocated by reasonable and otherwise trusted messengers, becomes doubtful to him.

Her book describing the process of making the documentary makes clear that it was a gruelling process for Anna, more so than for her counterpart. 'Doing this work is a battle,' she reflects. 'But I don't feel cut out to be a soldier. I am too soft; I get hurt easily; I take things personally.'[25] After I read all of her book and watched the documentary a second time, I texted Anna. Looking back after some years have passed, was it worth it? And how should we answer when people ask if they should argue with a climate sceptic? She responded immediately.

'Was it worth it? Yes. Should you argue with a climate denier? Only if you know there is an audience watching you can persuade.'

Of all the emotions that climate change can provoke in people, I wonder if denial is the most understandable. We're all climate change deniers to some degree. We have to be, to live our lives. Accepting the climate change science means living a kind of dual life, one in which you plan for a safe and stable future (invest in your retirement fund, meet your mortgage repayments, purchase funeral insurance) at the same time as you accept the evidence that it may not be. George Marshall writes, 'I have learned to keep worry to one side: knowing that the threat is real, yet actively choosing not to feel it.'[26]

Recognising that we all indulge in denial does not let deniers, especially professional and powerful ones, off the hook. We fear climate change, and some of us are increasingly prepared to change in order to cushion its impact on our world. Climate deniers—well, they fear a world transformed by action on climate change more than climate change itself. Understandable? Yes. Defensible? No.

CHAPTER 8

DESPAIR

Or the support group at the
end of the world

How many IPCC reports does it take to make a climate scientist cry? I guess about three or four. Or maybe, today, even fewer. One might be enough for them to burst into tears. Over the last few years, more and more climate scientists around the world have been 'coming out' about their climate-related depression, anxiety and anger. Some speak of uncontrollable sobbing or rage, not being able to get out of bed, abandoning research projects, deciding against having children or just walking around with the kind of 'existential whiplash' you get when you have a deep knowledge of what's happening but everyone else is carrying on as if it's business as usual.

Nobel Prize–winning professor Camille Parmesan, in a 2012 report for the US National Wildlife Federation, stated, 'I don't know of a single scientist that's not having an emotional reaction to what is being lost.'[1] Google 'climate scientist' and 'depression' and you'll find a stream of stories from people with PhDs and PTSD, what's known as 'pre-traumatic stress disorder', the mental anguish caused by preparing for the worst before it actually happens.[2]

For someone like me, who isn't a climate scientist, this kind of despair, while not as acute, is familiar. In preparation for writing this book, I've had to get across the science as much as possible, and I've read every climate-related story I've seen or been sent by friends and colleagues. And it's had a significant effect on my psyche. After having absorbed the science, well, there's no going back.

David Wallace-Wells refers to this as 'toxic knowledge'. Once you know about it, you can't unknow it, and your every thought is tainted by it.[3] I would be lying to you if I said I haven't experienced profound moments of despair and wild thoughts, including everything from thinking about moving houses or countries, to quitting my job, cutting off all my hair, learning to fire a gun and just lying down on the couch and not getting up. (And, of course, not finishing this book.)

All the writers and thinkers I am drawn to on this topic acknowledge that unhinged despair, if only fleetingly or in episodes, is an inevitable consequence of a thorough understanding of climate change, its causes and effects. The climate message is not just unsettling, it almost feels *disarranging*—not dissimilar to life during the COVID-19 pandemic. The idea that we're losing a common, liveable world, knowingly and deliberately through our own actions, is enough to drive some of us crazy. 'It's hard not to wonder what effect the news we hear every day about the state of the planet has on our mental state,' writes French philosopher Bruno Latour. 'How can we not feel inwardly undone by the anxiety of not knowing how to respond?'[4]

The area of climate change and mental health is big and getting bigger, with growing interest from governments, NGOs, public health professionals and researchers. Climate-related mental health problems take a variety of forms. Mental anguish and trauma caused by climate-related weather events or climate-related conflicts for the people who have lost their loved ones, homes and livelihoods. Depression and anxiety for farming communities caught in drought supercharged by climate change, not to mention those dealing with issues of food security and water scarcity.

Then there's the emerging research looking at whether temperature rises will cause more mental health issues, in particular aggression. For example, LA-based researchers conducted an individual-level longitudinal study of young people aged nine to eighteen and found that aggressive behaviours increased with rising average temperatures.[5] These associations were slightly stronger among girls and those from poorer families. (But, interestingly, the more green space a neighbourhood had, the less aggression was associated with temperature rises.) Other researchers have come to similar findings with adult populations.

Nick Obradovich from the Max Planck Institute for Human Development has conducted research showing that intense swings in temperature, a steady warming trend and the uptick in extreme weather events can cause small but cumulatively significant impacts on human mood and behaviour.[6] He researched how bad weather worsened the sentiments in people's social media posts (get ready for a more poisonous Twitter feed with a 2 degree Celsius temperature increase). He also found that if you increase the heat, even a little, in a complex system of human interactions, you will see the effects everywhere. These 'micro-scale climate effects' will undoubtedly impact our daily lives and wellbeing, with the potential to erode relationships and social cohesion.

The role of climate change in creating, exacerbating and growing mental health problems is well recognised by those in the public health and psychology professions. This has led various organisations representing mental health workers to speak out about climate change. The American Psychological Association has stated that the 'cumulative and interacting psychosocial effects of climate change ... are likely to be profound'.[7] Equivalent associations around the world have made similar statements. To address this new and rising need, they have developed reports and resources aimed at helping people manage their mental health in the climate age. And more and more psychologist groups, networks, alliances and specialist practices have sprung up to deal specifically with climate-related mental illness.

The potential for climate change to, as Latour puts it, 'make us feel inwardly undone'[8] has led to a long list of new terms to describe the state of mind shaped directly by knowledge of climate change: 'climate depression', 'environmental grief' or, the most popular, 'eco-anxiety'. In their work on the psychological impacts of global climate change, UK-based psychologists Thomas J. Doherty and Susan Clayton describe the symptoms of 'eco-anxiety' as including panic attacks, loss of appetite, irritability, weakness and sleeplessness.[9] They comment that these are similar symptoms to those suffered by people living close to hazardous waste sites. They go on to state that it can be difficult to differentiate between normal and 'pathological' anxiety about climate impacts.

A disabling worry about the health risks of an unsafe environment has, in the past, been seen by the mental health profession as obsessive behaviour, out of proportion to the reality of the threat. But as Doherty and Clayton rightfully point out, when it comes to climate change, can we really say the worry is irrational rather than rational? 'Given the evidence and predictions about health impacts of climate change and the unprecedented scale of those impacts, what constitutes an appropriate level of worry remains in question.'[10] The fear you have isn't a figment of your imagination, it's there in black and white in the latest IPCC report. This also leads us to question whether conventional psychology and psychiatry are well enough equipped to help people struggling with the myriad of climate-related mental health issues.

The despair of climate scientists is slowly and surely being shared by more people. There are signs of climate despair all around if you choose to look for them. Some of them are dramatic, like the lawyer turned environmentalist David Buckel, who in 2018 doused himself in gasoline and set himself alight in a park in New York. He wrote in his suicide note: 'Most humans on the planet now breathe air made unhealthy by fossil fuels, and many die early deaths as a result. [M]y early death by fossil fuel reflects what we are doing to ourselves.'[11]

Then there is the BirthStriker movement, people who have publicly declared they won't have children 'due to the severity of the ecological crisis and the current inaction of governing forces in the face of this existential threat'.[12] Look further down the rabbit hole of the internet and you'll find plenty of climate fatalism or eco-nihilism.[13] And plenty of what are called 'doomer' groups, from the far left of the political spectrum, who make Extinction Rebellion look like a collection of old-school Disney princesses (think Snow White rather than Mulan). They believe none of us will survive.

We can easily regard these stories as overreaction from overemotional or already disturbed people. Or just a new manifestation of age-old millenarianism in our midst. But the impact of climate change on the mental health of children and young people is less easy to dismiss. When researchers ask young people which are the biggest issues facing their generation, mental health often rates at the top of the top five (along with climate change).[14] In countries like Australia, the United Kingdom and the United States, there's an anxiety epidemic among young people.[15] Government bodies in affluent countries across the globe report increases in anxiety disorders in children as young as five, as well as an increase among young adults.[16]

There are theories that this rise in anxiety and growing understanding of climate change are linked; they may well be. What we do know is that researchers looking at the psychological effects of climate change on children have found both direct and flow-on effects from climate knowledge that place children at risk of mental health issues. The children in one Melbourne-based study showed high levels of concern about the climate, the mental health consequences of which could include 'PTSD, depression, anxiety, phobias, sleep disorders, attachment disorders, and substance abuse [which] can lead to problems with emotion regulation, cognition, learning behavior, language development, and academic performance'.[17] Our children seem to be inheriting a laundry list of mental health problems to go with a deteriorating planet.

More than simply making children anxious about the future or increasing adolescent depression, climate change is fundamentally altering young people's sense of 'self, identity and existence', according to academic Blanche Verlie from RMIT. (Her research is with Australian university students learning about climate change and involves those already engaged and concerned.) She has found that these students feel caught between knowledge about climate change and the pressure to ignore it and strive for individual success within a society driven by fossil fuel consumption. Extrapolating this research to the striking schoolchildren, she writes:

> climate change challenges the beliefs that ... if you work hard, you will have a bright future [and] adults generally have children's best interests at heart and can or will act in accordance with that ... Striking students ... are deeply anguished about what a business-as-usual future might hold for them and others. [Their] signs proclaim 'no graduation on a dead planet' ... This is not hyperbole

DESPAIR 157

but a genuine engagement with what climate change means for their lives, as well as their deaths.[18]

Verlie writes that because of this, young people may be the most susceptible to the negative effects of despair. When seen in this light, Greta Thunberg's statement at the United Nations that 'you have stolen our future' seems pretty reasonable.

Of all the emotions I've explored in this book, despair is undoubtedly viewed as the most unequivocally negative, both in terms of public perception and clinical opinion. Despair is the emotion or feeling of hopelessness that a situation is both profoundly wrong and will not change for the better. Despair is considered one of the most destructive of human emotions, associated not just with entrenched depression but with suicide.

One of the issues at the heart of despair is control, or lack of it. We despair because we experience low levels of agency or control, either actual or perceived. We're in a horrible situation and we're unable to act in any way to make it better. Children and adolescents experience lower levels of control over their individual and collective actions than adults, which can make them particularly vulnerable to experiencing despair.

We despair when the problem we face feels unrelentingly overwhelming: being in an abusive relationship, living in a refugee camp, dealing with a serious illness, living through a viral pandemic or some other situation where the exit strategy seems non-existent or beyond our capabilities. But climate change is a problem even the most optimistic of us finds oppressive in its magnitude. The enormity of it takes your breath away.

I suffer a bit from vertigo. I hate those elevators made of glass where you can see the view of a city as you shoot up to the higher floors of a building. If I'm sitting high up in a stadium, I find myself slumping in my seat and focusing hard on the ground. Thinking deeply about climate change, I experience a profound sense of emotional vertigo. I can't take it all in at once. I feel compelled to look away, to focus on a smaller part of it to cope, to search for something, anything, positive to head off despair and panic raising their ugly heads. As David Wallace-Wells observes, it's incredibly hard for us to pull the whole picture of climate change into focus, to look the science squarely in the face.[19]

The lack of control people in despair feel is amplified a thousand times when it comes to climate change. How do we control an invisible, odourless gas that is being generated by our lifestyles? How do we influence the way people in other countries like India and China and America and Saudi Arabia live their lives? And, more than this, how do we encourage the countless venal politicians and greedy corporations to make decisions based not on short-term profits and self-interest but on the long-term interests of 'ordinary people'? How in control can you feel if you live in a country without a democratically elected government or where the election process is corrupt, surrounded by voter intimidation and violence? Even in Australia, where more than 90 per cent of people vote in fair and free elections, their trust in their vote being important and influential has sharply declined in the last decade.[20] In a safe, wealthy and democratic country, we still despair.

Another aspect of climate despair sets it apart from your everyday garden-variety despair. Namely, it makes us lose our faith not necessarily in ourselves but in our fellow human beings. In fact, countless research projects have found that the more pessimistic you are about humanity, the less optimistic you are about finding ways to deal with climate change (and, of course, vice versa).

Climate action requires a Herculean feat of social cooperation at the local, national and international level. Even watching the unprecedented global response to COVID-19, such coordinated collaboration seems incredibly daunting for messy, inconsistent, arrogant and reckless human beings. And so climate despair can make people turn away from collective action and other people.

In an insightful piece of journalism on climate doomer groups around the world, ABC journalist James Purtill writes that the members of these groups can be distinguished from activists in groups like Extinction Rebellion by their tendency to turn away from society and possible action on climate change.

> Unlike rebellion . . . which looks outward, aiming to overthrow the established order in order to effect a better world, doomerism looks inward. Doomers speak a lot about their inner journey from despair and anger to acceptance. This makes sense; in the absence of political change, personal development is all that's left . . . They were planning to retire in the future and instead they

decided to retire now . . . And many of them pretty much stopped voting and participating in politics.[21]

The enormity of the issue, the lack of control and the feelings of alienation from other human beings that climate change despair can elicit in us lead some to conclude that it is an impossible problem. And many thinkers and writers looking at climate change within any discipline—economics, politics, diplomacy, philosophy and psychology—point out that it's a problem that strongly resists, even defies, human capabilities to solve it.

Professor Daniel Kahneman, author of the bestselling and highly influential *Thinking, Fast and Slow*, told George Marshall that he could see no path to success on climate change. You could almost say Kahneman is in despair. His theorised human brain, which combines short-term and long-term thinking, intuitive mental shortcuts and more complex decision-making, cannot solve climate change because of the way it responds to the threat of loss (more on loss later in the book). 'No amount of psychological awareness will overcome people's reluctance to lower their standard of living,' he told Marshall.[22]

But in response Marshall thoroughly questions this idea that climate change is an impossible problem:

> We are lucky that climate change is occurring now . . . when we have the combination of technology, wealth, education, and international cooperation that might be able to respond to it . . . What is more, the countries causing the climate

change will also be impacted by it [so] one positive result is the increased likelihood of action by these countries.[23]

In other words, if we can tap into the self-interest and fear of loss in affluent countries (two driving forces for human action that Kahneman writes about extensively), we might have a chance. Fundamentally, describing climate change as an impossible problem is itself a problem.

My other beef with the idea that climate change is an impossible problem is that it ignores important precedents, times when communities faced the prospect of being wiped out and yet still worked together for survival. Look at the endurance of Indigenous communities around the world, including in Australia, which have maintained their continuous cultures despite the brutality of colonisation. And at the global response to the COVID-19 pandemic, still unfolding as this book went to press. It makes me wonder if those framing climate change as an impossible problem are only doing so because *they* haven't managed to solve it.

Miranda Massie, the CEO of the Climate Museum, told me that while she understands personal moments of distress and despair, giving up is ethically unacceptable. 'It's a position almost exclusively taken in the United States by extremely privileged literary white men,' she said to me. She is of course thinking of the explosive September 2019 *New Yorker* article by writer Jonathan Franzen, which argued we have no chance of averting catastrophic climate change, and we should just admit this. In 'What if we stopped pretending?' he posits that those progressives fighting for environmental action are the ones in denial, that the so-called war on climate change is unwinnable. He argues that in order to prepare for the evitable, we need to admit defeat. 'False hopes are more dangerous than fears,' as Tolkien once wrote. While it would be unfair to characterise Franzen as an unalloyed doomer, his piece was still roundly criticised by climate scientists as inaccurate and unhelpful.

Again, this makes me wonder whether climate deniers and doomers don't have a few things in common—namely, an expectation of certainty and comfort; and an aversion to, maybe even contempt for, people involved in collective action trying hard, working against the odds, to change things. You can make a jump from denial to despair very quickly if you think this way. And don't be surprised if we don't see this increasingly as the years go on, with prominent deniers finally accepting the science but then saying, 'Well, it's too late now.'

Beyond this unique manifestation of white privilege in the despairing tone of some climate writing, the problem with the 'giving up' argument is that it rests on some unhelpful ways of looking at the world. The first is that this kind of despair, this doomerism, rests on the idea of certainty. In some respects, this is much like the kind of climate denial that insists climate change isn't happening and the future will be much like the present. Granted, despair about climate change rests on an understanding of the science, but it willingly ignores the fact that the future is not locked in.

US philosopher Terry Patten writes that the notion of inevitable doom at the heart of climate despair can often function as a self-fulfilling prophecy. 'The future is emergent,' he writes in an article for the magazine *Dumbo Feather*. 'That means we are creating it every day.'[24] Despair of this kind also relies on a narrow either/or way of looking at future. The world will either stay the same or become a living hell for everyone. We need to, as environmental psychologist Raymond De Young writes, break away from 'current folk mythologies of growth or apocalypse' and instead think about possible 'futures', plural.[25] Yes, all of these will be impacted detrimentally by climate change, but not all of these futures are the hellscapes of doomers' online projections.

You probably now expect every chapter in this book to try to demonstrate that even the most supposedly 'negative' emotional response to climate change has an upside, a benefit, however small. You'd be right. And despair is no exception. It's important to imagine the end of the world because it makes you reflect on the world we have and what you want to do to preserve what matters to you most.

David Wallace-Wells justifies the extreme portraits in his book *The Uninhabitable Earth* on the basis that it makes sense to prepare for the worst. He argues that in dismissing the worst-case scenario, we become complacent (better to be over-prepared).[26] Furthermore, theologian Hannah Malcolm writes that in countless religious traditions, descriptions of the end of the world have two clear functions for believers: to make room for repentance before it's too late, and to offer the chance, however slim, for renewal, a pathway towards a better world.[27]

In light of this, it's worth considering a less nihilistic and potentially constructive version of despair than the one offered by hardcore doomers. It's called Deep Adaptation. It started when in mid-2018 Professor Jem Bendell from the University of Cumbria in the United Kingdom wrote an occasional paper entitled 'Deep Adaptation: a map for navigating climate tragedy'.[28] Bendell is a well-respected academic in the area of sustainability, and more than anything the article feels like a repudiation of his own discipline. He is scathing of the way sustainability studies have tried to focus on positive solutions to climate change in the form of mitigation strategies and golden opportunities provided by technological innovation. He argues that much of the bright-siding and highlighting positive information about progress when it comes to climate change is close to useless. 'Discussing progress in the health and safety policies of the White Star Line with the captain of the *Titanic* as it sank into the icy waters of the North Atlantic would not be a sensible use of time,' he comments dryly.

Instead of seeking balance or worrying about triggering possible despair in his audience, Bendell starts his argument from the premise that the end of the world is, in fact, nigh. He is explicit upfront that the aim of the article is to provide readers with an opportunity to reassess their lives in the face of what he calls 'an inevitable near-term social collapse due to climate change'. He goes through some of the climate science to justify this judgement of inevitability, but takes the worst-case outcome for granted and then asks the reader, what next?

It is difficult to predict future impacts. But it is more difficult not to predict them . . . We do not know what the future will be. But we can see trends. We do not know if the power of human ingenuity will help sufficiently to change the environmental trajectory we are on . . . We might pray for time. But the evidence before us suggests that we are set for disruptive and uncontrollable levels of climate change, bringing starvation, destruction, migration, disease and war . . . But when I say starvation, destruction, migration, disease and war, I mean in your own life. With the power down, soon you wouldn't have water coming out of your tap. You will depend on your neighbours for food and some warmth. You will become malnourished. You won't know whether to stay or go. You will fear being violently killed before starving to death.[29]

Bendell makes a compelling case that contrary to popular wisdom, despair can actually be helpful. 'The loss of a capability, a loved one or a way of life, or the receipt of a terminal diagnosis have all been reported, or personally experienced, as a trigger for a new way of perceiving self and world, with hopelessness and despair being a necessary step in the process.' He points to the fact that inviting people to get together and consider collapse has forged new and creative support environments, with people helping each other to imagine a different kind of collapsed world together.

Finally, he offers up what he calls the 'Deep Adaptation Agenda', which is less a plan about how to survive climate chaos by drinking your own urine and more a series of questions we need to ask ourselves as we prepare for what's coming. 'What are the valued norms and behaviours that human societies will wish to maintain as they seek to survive? What do we need to let go of in order to not make matters worse? What can we bring back to help us with the coming difficulties and tragedies?'[30] The kinds of strategies he is talking about here are extreme to imagine for the kinds of entitled, affluent societies people like me live in: a complete overhaul of our lifestyles, the geography of our cities, and how we work and play, travel and eat.

For a 35-page academic article with lots of fancy jargon, it has reached well beyond the ivory tower and been downloaded hundreds of thousands of times. What's more, Deep Adaptation groups have popped up around the globe and Bendell's work has been embraced by some in the Extinction Rebellion movement. In environmental circles, his work has provoked enthusiastic discussions and disagreement.

I am yet to be convinced that descriptions of 'inevitable near-term social collapse' are a great starting point for conversation with the kinds of people I have in my focus groups, people who need to be engaged and convinced. But as I said, imagining the worst, even if only every now and then, can inspire us to do more and reflect on what kind of life we want to lead in an emerging climate-affected world. We do need to ask ourselves the question Bendell poses: Supposing it's too late, what then? But we also need to ask ourselves the question: Supposing it's not too late, then what?

While I don't believe near-term social collapse is inevitable, I do know for sure that greater and more complex mental health problems associated with climate change will emerge around the world. Already there are more and more climate anxiety support groups and climate grief workshops to be found both online and in person. While they take different forms, all are driven by a desire to stay engaged with other people in the community while grappling with feelings of frustration and hopelessness. There's even a version of the Alcoholics Anonymous step program for climate grief run by the Good Grief Network based in Arizona, the aim of which is to build personal resilience and strengthen community ties to combat despair and eco-anxiety.

Finding emotional resilience through community is what compelled Cassia Read to start her own climate support group. I first met Cassia in the courtyard of Castlemaine Gaol, no longer a place for outlaws but for arts festivals and community events in regional Victoria, Australia. She was sitting on a blanket surrounded by string, paint and squares of cloth cut from second-hand sheets, with her sister and children by her side. She was inviting people to draw or paint on the squares, to visualise their response to climate change, their hopes and fears about the earth. These squares would then be strung up around the town and in other places as Climate Flags.

The project was inspired by the Tibetan practice of prayer flags, colourful cloth banners hung across temples and other buildings, used to bless the surrounding countryside. Before I met

Cassia I had seen the flags hanging around the town and in shop windows. She invited me to create a flag of my own. I drew a heart in different colours and added some words about wanting to protect my kids by doing something about climate change. Chatting as I drew, we found out we had some mutual friends in town. I was interested in both the Climate Flags project and the group that creates them as a form of collective therapy to address climate despair.

Cassia grew up in a family of nature lovers and artists. She studied ecology at university and went on to work at the Royal Botanic Gardens in Melbourne creating Fungimap, a big citizen-science mapping project for understanding fungi in Australia. She then did a PhD in the Mallee and the Wimmera, very remote landscapes in her home state of Victoria, looking at the moss and lichens carpeting the ground in those areas and their critical role in the ecosystem.

But her interest in climate change pre-dates her studies.

I think it started back as a teenager. My brother's nine years older and was involved in environmental campaigns and the Greens Party. He talked to me about climate change way back in 1987. I used to lie in bed awake stressing about the future and the environment. The environment has always been my solace and refuge in times of hardship, and this sense of impending doom in the environment was even more challenging for me, because there is no refuge or retreat.

Cassia is one of those scientists who approaches climate change with an intimate knowledge of how nature works. She describes herself as non-confrontational and 'not an activist'. 'What I'm good at is quieter, more reflective science-y stuff, gently teaching people about nature.'

And yet a range of personal, professional and political events increased her level of worry and determination to do something public and loud with her increasing eco-anxiety. She had children. She watched as Australia's political leaders twisted and turned and failed to come to grips with climate change policy (Australia is the only country that has ever introduced a tax on carbon and then repealed it). And she authored a report for Parks Victoria on the impact of climate change on national parks, which predicted some terrible damage being done to their ecosystems.

> My stress levels started going up and up. I spent a lot of nights worrying about the future and feeling really helpless. I wanted to come out of the paralysis but I wasn't sure how. I spoke a lot with my sister who was similarly stressed about it, and over the years we kind of brainstormed. 'What can we do? Write letters? Take to the streets?' It was hard to know what to do.

The idea for Climate Flags grew out of this desire—hers and her sister's—to combat their feelings of helplessness. They imagined letters to politicians would be ignored, so instead they thought about prayers written on fabric (more permanent) and hung up for all to see (more public). Once they had the concept, the next step was gathering people around, not only to create the flags, but to have difficult but necessary conversations about climate change.

> I guess I was just really looking for a way to engage the people around me. Wake them up gently without traumatising them . . . How can we create a bridge for the people around me, my dear friends who don't want to talk about climate change, who don't want to engage but they love nature and they've got children?

Cassia started hosting regular kitchen-table conversations with people, mostly parents, in her community, people who weren't already engaged in climate activism but 'loved gathering and doing something creative'. 'It was quite powerful in our community,' she told me. 'There were moments where we were sitting around and sewing and there were some really painful, raw, real conversations about climate change. People spoke about how they were feeling and wondered what they could do, and other people would give them ideas.'

Since Climate Flags launched in early 2019, thousands of flags have been created and events held in and around Castlemaine and Melbourne. Making the flags in public places like markets and festivals sparks a lot of difficult and emotional conversations, Cassia finds. 'You almost feel like you need to be a climate counsellor to do this. People just want to tell you about the grief or anxiety or hopelessness. I feel like there's not the psychology to deal with this unprecedented situation.'

I ask her what kind of flags different people create, and she tells me there's an amazing range, from the importance of tiny houses and green energy to wanting to protect children and grandchildren from a dire future.

> A lot of people just want to express a love of the earth and don't want to do anything really depressing or angry because they're aware it's going up in the community. They want it to be uplifting. No one's written 'We are fucked'. Only teenagers. But they say that quite happily, with the cheeriest smiles on their faces.

The most meaningful moments with the project, for Cassia, have been when large groups of people have come together to create the flags, like at their launch, which attracted 100 people from the area. 'Everybody who came just felt it was really profound to sit with a lot of people in a space where we were all acknowledging the gravity of the situation.'

While her kitchen-table group hasn't read and discussed Jem Bendell's paper—too confronting for the Climate Flag crew, she tells me—she agrees that despair can be a necessary and helpful emotion to feel, if only for short periods of time. 'There's no way around that feeling,' she reflects. 'You can't just jump to taking action without going through that grief and anxiety.'

Starting Climate Flags with her sister has allowed Cassia to manage her personal response to the climate crisis. 'The good news for me out of this whole experience is my anxiety levels have gone down. Rather than lying in bed at night thinking about rainfall in twenty years, I think, "I'm going to have this conversation with someone" or "We're going to make flags" or "What else could we do?" I feel like I can be part of the solution.'

In a fascinating study of schools, teachers and students in North Carolina, researchers found that the more individuals despaired about the climate, the less likely they were to engage with pro-environmental activity. But they also found something interesting about the interaction between hope and despair—there was little relationship or interaction between the two. Students who were environmentally active could feel hopeful but also express feelings of despair as well as high levels of concern. 'Directly acknowledging and addressing feelings of despair may be an important part of avoiding . . . inaction among adolescents,' they concluded.[31]

As Cassia Read points out, despair about the climate is not something you can bypass or resist. Like anger or fear, it's a place you find yourself returning to from time to time, although it's not somewhere you want to set up house. Professor Tim Flannery put it to me bluntly in these terms: 'Despair is the lazy man's option.' (He's much more interested in rebellion.) And if the behaviour of doomers is any guide, despair leads to inaction, which in turn narrows the range of possible futures we can hope to create.

Feeling despair every day? Here's how to get help . . .

I can't provide expert advice because I'm not a mental health professional, but in the process of writing this book I've come across numerous resources and organisations that can help if you're finding yourself overwhelmed with despair on a regular basis in a way that's negatively impacting your life and the life of those around you.

Some of those resources are listed at the end of this book, while the book's Conclusion gives you a few principles I've developed for managing my own mental health and energy in order to continue working on climate-related issues.

One useful tool when thinking about managing all the disruptive and difficult emotions climate change can trigger is the 'window of tolerance'. First named by American psychiatrist Dan Siegel, it involves creating your own personal boundaries—or 'zone'—where you're able to function most effectively, so that you can process information without too much mental anguish. It helps you to avoid extremes, neither shutting out all information that's distressing nor going the other way, and prevents you endlessly consuming upsetting information and then falling into despair.

Interestingly, even though they don't use this term, many of the activists I interviewed for this book talked about monitoring and controlling the amount of upsetting information they consume in order to continue to work on climate change. That might be not engaging with social media, or only reading some, not all of the new climate science.

Consider the following options as well:

- Find friends who share not only your concerns but also your commitment to hope and action; catch up with them regularly for something energetic and serotonin-generating, like a long walk.

- Seek out mental health advice from a professional who is connected to or sympathetic with the various ecopsychology and climate grief networks.
- Find ways to talk about this with the people around you, even if they don't completely share your views, and tell them how you're feeling.

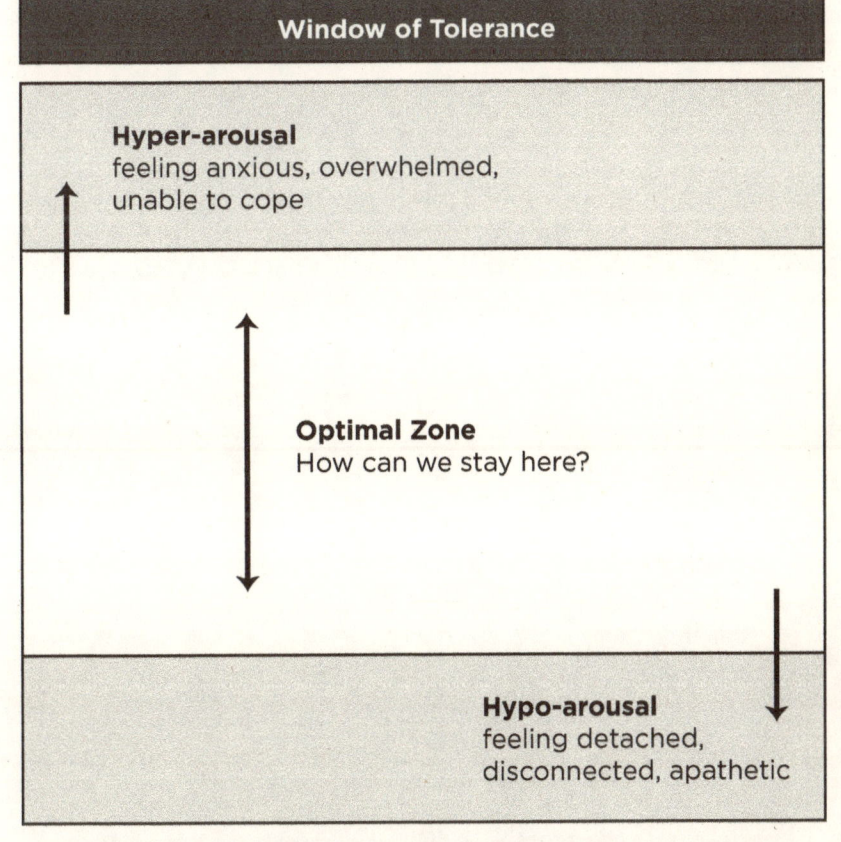

Window of Tolerance

Hyper-arousal
feeling anxious, overwhelmed, unable to cope

Optimal Zone
How can we stay here?

Hypo-arousal
feeling detached, disconnected, apathetic

First named by Dan Siegel, the window of tolerance is a concept that shows how you can operate in a 'comfort zone' and avoid hyperarousal or hypo-arousal, both of which can make you shut off or shut down.

DESPAIR

CHAPTER 9

HOPE

Or how to get out of bed
in the morning

When I think about hope, I think about Chido Govera. I first met Chido in the green room of the Sydney Opera House, where we were both speaking at an event run by the global organisation MAD. Through its workshops, conferences and events featuring people in the restaurant industry, farmers, thinkers, writers and activists, MAD aims to create a 'better, healthier, more sustainable, more delicious world for cooks and eaters alike'.[1]

Food issues related to climate change are regularly discussed at MAD events like the one I was part of in Sydney. The organisation was founded by world-renowned Danish chef René Redzepi, who has managed to pull into the MAD tent other gurus of the restaurant world, including Massimo Bottura and David Chang. And indeed, on that day in Sydney, all of those chef superstars were pacing the room, along with famous Australian chefs, like Neil Perry and Kylie Kwong. I was somewhat bemused to be there and felt like a sustainably farmed salmon out of water. I'm not a chef, restaurant critic or professional foodie. I was only there on the invitation of the head of Talks & Ideas at the Opera House, who suggested to the MAD organisers it might be good to include a perspective on how 'ordinary' Australians feel about food (one of my research areas of interest).

Of all the speakers I knew would be at the MAD event, Chido was the one I wanted to meet the most. Her CV read like an against-the-odds movie script. Born in 1986 in Zimbabwe, she was orphaned at the age of seven when her mother died of AIDS. She never knew her father. After her mother's death, she lived with and cared for her 100-year-old grandmother and her younger brother. She endured physical and sexual abuse at the hands of family members and had to leave school at the age of nine to work full time, digging in people's fields all day to get food for her family.

When Chido was ten, her cousin (the daughter of one of her mother's sisters) suggested marriage to a man 30 years her senior, but Chido resisted. (Approximately one in three girls in Zimbabwe are married before they turn eighteen.) In 1998, at the age of eleven, with the help of a woman from a local church, she enrolled in a week-long program at Africa University, financed through environmental entrepreneur Gunter Pauli's Zero Emissions Research and Initiatives (ZERI) Foundation. There she learned how to grow oyster mushrooms using cornstalk waste (just an aside, mushrooms are an important climate-era crop given they can be grown indoors). After working in the university's lab between the ages of twelve and sixteen, she was not only able to feed her family, but to teach other orphans to grow mushrooms successfully.

Since then, through the organisation she founded, the Future of Hope, Chido has taught mushroom culture to people from her native Zimbabwe, as well as communities in other parts of Africa, and in India, China and parts of Europe. She has pioneered new techniques, such as growing mushrooms from coffee grounds, even training a group of hipsters in San Francisco. She now lives on a farm, once an orphanage, in a suburb north of Harare, where she tends an extensive garden, looks after the orphans she inherited with the farm, runs her foundation and grows mushrooms to sell at markets around the city.

During a lull in green room conversation, I crept up to Chido to introduce myself. 'What are you going to talk about?' I asked her. 'I don't know,' she replied. 'I will listen to what other people say before me and I will decide.' I looked around at the seasoned media performers and TV chefs in the room as they paced around nervously with notes, anxious about addressing the sellout crowd. When Chido got up, luminous in a short-sleeved white and silver dress with her dreadlocks pinned away from her face, she spoke without notes and captivated the hushed audience. The theme of the day was 'tomorrow's meal', and she addressed the topic with warmth and wisdom.

'Tomorrow's meal', she said, has to be a key to unlock the potential of all people, a product of real collaboration, something that creates opportunities for women and girls who have been denied food, freedom and education. 'A driver of social and economic change . . . A peacemaker that unifies us all.' After her talk, as we gathered for a meal in one of the Opera House's fancier restaurants, I approached her again to tell her how impressed I was by her work. 'What can I do to help?' I asked. 'I'd like to come back to Australia and bring with me some of the girls I work with,' she replied. Thanks to the support of MAD, her Australian tour happened a year later, and Chido and I have been friends ever since.

I spoke to Chido just after Christmas 2019, me sweltering in my study, her rugged up in a friend's lounge room in snowy Belgium. She was taking some much needed rest after a hectic year of work and family responsibilities. We talked about the impact of increased warming and weather events in her country. Zimbabwe might be landlocked, but climate change is a pressing concern, with droughts affecting rural and urban water supplies and food security. More frequent and more severe cyclones and floods are damaging property and infrastructure. The situation is made worse by the other problems Zimbabwe faces. It might be rich in minerals and natural assets, but a corrupt government, with waves of civil unrest, means there is widespread poverty and deprivation as well as sky-high inflation and significant public health issues like HIV/AIDS and tuberculosis. Communities don't bounce back after extreme weather events as they do in countries like Australia.

I asked Chido if she had noticed the signs of climate change in her own work as a farmer. 'Absolutely,' she said. 'There are more droughts and changes in the rains. You can read it in the way the crops grow. I used to say it has changed since I was a little girl but actually we have all noticed the changes in the last seven years.' As she travels around Zimbabwe and other parts of Africa, the communities she works with tell her they too see the signs of deterioration. 'Talking to them about the science isn't necessary,' Chido explains. 'They see the changes and so what you have to do is teach them the skills that will help them adapt, which includes growing and eating different foods.'

Chido believes adaptation is possible, even in poor communities like the ones she works in. For example, a staple food in Zimbabwe is sadza, a maize-meal porridge, sometimes served with vegetables and (if you're lucky) meat. Sadza is usually made with corn (hence the surplus of cornstalk waste for growing mushrooms), but as Chido tells me, once upon a time it was made with other grains that are hardier in the face of climate change. Along with cultivating mushrooms from waste to add essential nutrients to people's diets, Chido is trying to encourage communities to return to more traditional foods. 'If you explain the value to people they will actually change, they see we haven't always done things this way. We're preparing them for a future where we know things will be harder.'

Chido and the women and girls she trains are part of that massive global contingent of small-scale farmers Katharine Wilkinson spoke about in her TED talk (see Chapter 3), whose practices have the potential to reduce carbon emissions. Educating her girls and teaching them to farm go hand in hand for Chido. We have spoken too about the importance of including education about reproductive health in the work she does with communities.

In her interviews around the world, Chido is often asked, 'What does hope mean to you?' It was an irresistible question for me to pose too, given her background, given the extraordinary obstacles she has faced and overcome. Once upon a time, she looked like a girl without a future. How did she remain hopeful and not despair?

> I felt angry at my mother that she died. I know it sounds crazy but I did. So when she died I was determined that I wouldn't let my brother down, that I would find a way to feed and educate him so he wouldn't have to go through what I went through.

She was strong and self-assured enough to know marriage wasn't that chance for a better life, either for her or her family. When she was invited to learn how to cultivate mushrooms, she saw in that the pathway to achieve her goal and in turn became inspired to teach other orphans like her the same skills. Despite all the challenges associated with farming, running the foundation and being a foster mother, not to mention living in Zimbabwe, Chido continues to derive hope from watching what can be achieved when communities are given tools to help themselves.

> Hope is the commitment to acting on and trusting one's belief of a future different and better than the past or present. It is believing that change is possible while

being fully aware of whatever adversities one might be confronted with today and yet being so focused on investing all the effort necessary for a better and brighter future. In my work, giving hope is really just asking people what kind of future they would like to live in, giving them a say about that, then giving them the skills and tools to make that future.

Over the years I've noticed Chido is a fan of the inspirational quote. She posts them a lot on her Facebook page. 'Accept yourself, love yourself and keep moving forward.' 'Forgiveness is the merciful grace of compassion in action.' I tell her if she was anyone else I'd block these cheesy sayings but she gets away with it. They're mixed up with updates on the achievements of her girls and women of colour around the world, as well as news about Zimbabwe I don't get to hear about through my local news. But if I had to create my own inspirational quote meme to post online, it would be a line from Nathaniel Rich's book, *Losing Earth*: 'False hopes are worse than no hope at all.'[2] If we pin our hopes for a liveable future on some miraculous invention that will mean we can keep on living as we do and still stay under 2 degrees Celsius of warming, well, we're just dreaming . . .

There are as many different kinds of hope as there are grains to make sadza, and some are better suited to the climate age than others. Hope is always positioned as a positive emotion, the antidote to fear and despair. Hope is an expectation, a longing, a desire, a dream, even a daydream. Whether those hopes are based in reality or not is another question. Hope can never be a completely positive emotion, even if the act of hoping for something has positive benefits. The nature of hope is premised on the recognition, even faint, of the possibility of a negative outcome. In others words, hope is 'fearing the worst but yearning for the better'.[3]

Psychologists argue that we human beings are hardwired for hope by our 'optimism bias'. This bias leads us to believe that we're less likely to suffer misfortune and more likely to achieve success than reality would suggest. This phenomenon was initially described in 1980 by Neil Weinstein, a social scientist at Rutgers University in New Jersey. His study found that the majority of college students believed their chances of developing a drinking problem or getting divorced in the future were lower than those of their peers. This majority also believed that their chances of positive outcomes, like owning their own home and living into old age, were much higher than those of their peers.[4]

When thinking of the future, regardless of what the data shows, we overestimate the good and underestimate the bad. Many people think they won't catch a deadly virus, for example, even when it's extremely contagious. Of course, we wouldn't get married, have kids or start businesses in the first place without this optimism bias. Some of us would struggle to get out of bed in the morning if we calculated the odds of having a bad day. But you can easily see how this optimism bias manifests itself in thinking about climate change in various, not very helpful, ways.

For example, numerous studies have shown that while participants might agree climate change is real, is caused by humans and will result in sea level rises and extreme weather events, they're not as convinced that they personally will be affected by any of these things. Studies by Tony Leiserowitz and others have found that individuals generally consider climate change 'less serious' and 'less dangerous' to themselves than to others. Climate change is something that happens to other people.[5]

Further research has found that when individuals are reading information about climate change, such as an IPCC report, that uses the language of probability and uncertainty of outcome, they often misinterpret the intended messages in an overly optimistic manner.[6] In other words, at any hint of uncertainty they put a 'best-case scenario' spin on it. So the less definitive the conclusions, the more wriggle room there is for people to infer a shiny, happy outcome, even if that is unreasonable. That kind of optimism can reduce people's motivation to act because they underplay the seriousness of the situation. But expressing results with this kind of uncertainty is virtually obligatory in any serious scientific publication.

No wonder 'optimism bias' is also referred to as 'the illusion of invulnerability'. It's one of the reasons we can agree that climate change is a serious threat but don't change anything about our own lives. Climate change is often written and talked about in terms of what seem to be tiny changes, like a 1 degree Celsius temperature rise or 45 centimetre (18 inch) sea level rise. To the lay person these seem like bearable shifts in the environment, easy to adapt to. What hasn't been communicated well is that tiny changes can lead to big, cumulative effects.

The capacity for our optimism bias to breed wishful thinking on climate change is prodigious. The most insidious and alluring climate change wishful thinking involves the idea that technological innovation will solve our problems. That Elon Musk and Richard Branson will get together and develop a CO_2-sucking machine that we can attach to every car, plane, house and building and continue to live as we have over the last 50 years without worrying about a warming earth. (They might call it the Greta 2020.)

There are, in truth, some promising inventions out there that might help us in our urgent need to adapt to and mitigate climate change impacts. The whole area of geoengineering is exciting to read about, with everything from pumps for cooling coral reefs to carbon-capture concrete. But very few of the serious thinkers on climate change believe these or any inventions of the near future are going to 'solve' climate change for us.

The problem too is that the 'tech will save us' mindset has the capacity to make us more complacent about the main pathway to change: moving swiftly away from fossil fuels and ensuring widespread changes to human behaviour, especially in affluent countries. As George Marshall writes, 'turning up the optimism about technology has a capacity to turn down the volume of threat. It's only a few more notches on the dial before one is deep into outright denial.'[7]

Marshall's statement is supported by psychological studies. Victoria Campbell-Arvai and her colleagues from the University of Michigan conducted a survey of nearly 1000 American adults on their attitudes to carbon dioxide removal strategies, or CDRs.[8] They found that learning about these strategies reduced the perceived threat of climate change and indirectly reduced support for mitigation policies. This was particularly true for political conservatives; indeed, surveys have shown that conservatives are more likely than other groups to believe that technology, rather than government policy or behaviour change, is the solution to climate change. The researchers concluded that we should be careful when we promote technology as a fix for climate change, as it may undermine support for climate change action, particularly among the disengaged.

The good news is that despite our general obsession with technology, we generally don't see it as the climate silver bullet in this context. The 2015 Pew study found that a global median of 67 per cent say that in order to reduce the effects of climate change, people will have to make major changes in their lives. Just 22 per cent believe technology can solve this problem without requiring major behaviour changes.[9]

We can conclude from all this that there are downsides to hope if it encourages us to be overly optimistic and therefore less likely to act.

What does psychological research tell us about the influence of hope on believing in and acting on climate change? Not surprisingly, the results, again, are mixed. American academics P. Sol Hart and Lauren Feldman studied the impact of climate change news imagery and text on intentions to conserve energy and engage in political behaviour related to climate change. They found that hope was an important aspect of any effective messages, and that it went on to strengthen intentions for political participation.[10] But in another study published in the same year, Cornell researcher Hang Lu found that, in a direct comparison, sadness was more effective than hope in getting people to seek information about climate change and support certain pro-environmental policies.[11]

The difference in these findings regarding the capacity of hope to make us act could be driven by all manner of things. Lu was getting participants to respond with either hope or sadness in relation to a sea star wasting disease event in the North American Pacific region. Sol Hart and Lauren Feldman were showing participants images of solar panels and asking them their thoughts about green energy. What different participants were being asked to do—vote, sign a petition, seek out information, change behaviour around the home—in response to hope or sadness or anything else was also different. And herein lies the problem with hope: it can function as what psychologists term an 'unspecific action tendency'. Hope can be 'yearning for a positive outcome' but not much more than that.[12]

The thinkers, researchers and writers I admire tend to define hope in what you might call 'negative' ways. Defiant hope. Stubborn hope. Stoknes calls this 'active scepticism' or 'grounded hope'. 'Our situation is desperate and at the same time hopeful,' he writes, trying to capture the sentiment.[13] Writer Rebecca Solnit calls it 'hope in the dark'—less like hope and more akin to grit, determination and resolve.[14] In her interview in the magazine *Dumbo Feather*, academic Susanne Moser advocates thinking about hope as a form of 'functional denial':

> There are many different flavours of hope. One is sometimes called grounded hope, active hope, or authentic hope. That's where you are not at all convinced that there is a positive outcome at the end of your labours . . . But you do know you cannot live with yourself if you do not do everything towards a positive outcome. And then there's radical hope [where] you don't know at all whether the outcome is positive or negative . . . Between grounded hope and radical hope, that's what we're going to need for climate change.[15]

The best kind of hope for the climate change era means accepting the stark reality of what climate change has done and will do to our planet. It requires a certain comfort with uncertain outcomes. It requires a bloody-minded attitude to moving forward regardless. And, more than anything, it requires a belief in the power of the collective, or groups working together, to get things done.

The research on hope shows that collective or group action is critical to developing the kind of grounded hope or sceptical activism writers like Moser and Stoknes advocate. Jochen Kleres and Åsa Wettergren found that the young Swedish and Danish climate activists in their study generated active hope through working with others. 'Trust in "one's own" *collective* action seems to be the essence of the hope that activists talk about,' they wrote. This was in contrast to the hope that 'others will fix the problem', which activists identified as one of the big barriers to getting other people to act on climate change.[16]

The importance of the collective as a vehicle for generating hope goes beyond activist circles. Jumping back to comments in a previous chapter about evolutionary psychology, we have evolved over millennia as herd animals. Groups that were able to work together survived and thrived. Many of us in Western, affluent countries, with our focus on personal autonomy and individual rights, are prone to forget the powerful influence of the group. We dismiss it as peer pressure. Instead, as George Marshall writes, social conformity is not 'some preference or choice':

> It is a strong behavioural instinct that is built into our core psychology . . . It originated as a defence mechanism during our evolutionary development, when our survival depended entirely on the protection and security of our social group . . . There are, therefore, real and serious risks involved with holding views that are out of step with your social group.[17]

Given so many of our other evolutionary traits work against any action on climate change (short-termism and risk vividness, for example), it's not surprising that researchers have been keen to test the capacity of social conformity to encourage belief and action on the climate threat. And there have indeed been some promising results in this area. Researchers from Yale University have shown that perceived social consensus can reduce ideological biases on climate change. They looked at nine nationally representative surveys of Americans (amounting to more than 16,000 respondents) and found that ideological differences in climate change beliefs, attitudes and policy preferences are smaller when people's close friends and family members care about climate change. Being surrounded by people who care about climate change is more influential on our views than, say, the attitudes of people who vote the same way we do.[18]

What our tribe or the larger herd is actually doing can also be influential, not only on our views but our actions. In the 2001–02 California energy crisis, the government was effective in getting consumers to change their energy use, not just by appealing to their hip pocket, but also to their sense of civic pride. The 'Flex your power' campaign framed its messages—'Together we can do this', 'Help us all get through the power emergency'—to connect the actions of people 'just like them' to the powerful effect of people acting together.[19]

A now famous study by Norwegian researchers showed that hotel guests were more likely to re-use their towel for environmental reasons if they were told the vast majority of other guests were doing so. The 'combined normative' appeal (this is what other people are doing) was far more effective than a straightforward demand (do it), a descriptive approach (re-using towels helps the environment) or the standard hotel message (we would like it if you re-used towels to help the environment).[20]

More recently, messaging as the 2020 COVID-19 pandemic unfolded ignited an unprecedented change in behaviour around the world, with millions of people upending their daily routines, businesses rethinking entire systems, and the global scientific community dropping previously held norms of competition and privacy to pool information and resources. As Harvard Medical School professor Dr Ryan Carroll told *The New York Times*, 'The ability to work collaboratively, setting aside your personal progress, is occurring right now because it's a matter of survival.'[21]

So there is hope in the persuasive power of the human herd to influence both thought and deed. Working in groups, teams, communities of like-minded people, that's where we find rechargeable, renewable batteries for ongoing power generating hope and optimism. This is why it matters not just what we think about climate change, but what other people think. And why it matters, in terms of sustaining behaviour that helps the environment, that we see other people doing what we're doing and feel as if our actions are a small part of a larger picture. Otherwise anything we do in an attempt to make a difference can just seem futile.

―――

Ironically, writing this chapter about hope has not made me more hopeful about climate change. It has helped me work out what kinds of hope are helpful and what kinds of hope are, in truth, denial in disguise. There are the ludicrous hopes, for example, that climate change isn't happening or will in fact be beneficial (we can grow champagne grapes in southern England!

People will enjoy balmy summer vacations in Scotland!). There are the false hopes that the gurus of technology will develop new inventions at any moment to maintain the status quo without us having to do much more than install solar panels on our rooftops (thankfully, the Pew survey shows that the only countries that believe this to any degree are Saudi Arabia and Japan).[22]

Instead, writing about hope has helped me zero in on the kind of determined, resolute mindset I'm going to need to face the immediate future, a hope that is in fact sceptical activism. And that kind of hope can't be a vague dream for the future that requires no personal sacrifice or action. It has to be a deliberate choice. All the people I interviewed for this book say the same thing differently about hope, namely that you choose hope even in the face of the evidence because it's the right thing to do, both morally and practically. Anna Rose put it to me this way: 'I realised almost at the beginning of my career as an activist that hope is a strategic decision. No one will join a movement that doesn't have hope.' Hope is an ethical imperative. Action generates hope. And hope attracts others to your cause, which might, just might, start to shift the odds in our favour—and the planet's favour as well.

CHAPTER 10

LOSS

Or bury me in a carbon sink

After the wildfires that swept across Australia, starting as early as September 2019 and reaching their peak in December 2019 and January 2020, the country took some time to calculate the loss. The media reported astounding figures in terms of destruction of property, in the hundreds of millions of dollars. Then there was the loss of income and livelihood for those still trying to do business in fire-damaged areas. The damage to infrastructure, which would require significant taxpayer dollars to repair. The numbers of animals killed and wounded—impossible to calculate, but possibly over a billion—with iconic animals like the koala and the platypus now endangered or seriously threatened. Finally, there is no dollar amount we can attach to the loss of 33 human lives. But the gravity and breadth of the catastrophe meant something else had been lost, something Australians have grown to cherish as a national birthright: a relaxing summertime.

Summer in Australia is when we celebrate Christmas and New Year, along with our national holiday, and when we take a long school holiday or work break. Some of us go away to beach and bush houses a few hours' drive from our homes. It's a time for family and friends, swimming in salt water and ignoring our emails and smartphones. It's rarely a time for politics and serious business.

We'd had bushfires over summers in the past, of course, and opened our wallets to assist those affected, but the scale of those 'typical' summer fires had rarely been enough to stop a nation from pouring themselves a beer and tuning in to the cricket. That being said, in the last few years I've noticed a mood of anxiousness and concern creep into conversations about our summer in the focus groups I've conducted. Australia is experiencing hotter summers, so much so we keep breaking official records for the hottest summer year after year. The public has noticed. In my 2019 essay *Australia Fair, Listening to the Nation*, I wrote:

> Australians are slowly getting it. Our way of life is changing. It isn't just that you can't let the kids run through the sprinklers in the backyard anymore, or that there are fewer bugs on the windscreen after a night-time drive. The Australian summer—traditionally a time of best-selling novels and drinks, family and friends—is being transformed into a time of tension and worry. There is anxiety about unprecedented weather events threatening houses and food chains, record temperatures taxing the health of older Australians, family pets and young children and sky-high electricity bills putting pressure on household budgets.[1]

I didn't predict the kind of wide-scale disaster that happened in 2019–20, but if I had been paying more attention, I would have realised that it was coming. Respected economist Professor Ross Garnaut wrote in 2008, in his government-commissioned *Climate Change Review*, that without adequate action, the nation would face a more frequent and intense fire season by 2020.[2]

In the midst of so much tangible loss, we've lost something intangible but equally precious.

The notion of loss comes up time and again in the literature on climate change. In terms of emotion, loss is a step beyond sadness towards grief, anguish and pain caused by losing someone or something of high value. But at a more basic level, loss can refer to having to (or being forced to) give something up, a process psychologists have been particularly interested in for some time.

Probably the most important concept when it comes to loss in psychology is 'prospect theory'. Developed by Daniel Kahneman and Amos Tversky, it relates to the notion of loss aversion, namely that humans are more sensitive to losses than to gains. The two authors explained in their 1979 article on the topic that people underestimate outcomes that are probable in comparison with those that are certain, a tendency that drives risk aversion in our choices.[3] Put simply, we're highly sensitive to losses in the present, which feel more certain, than losses in the future, which feel less certain. Furthermore, when it comes to gains, we prefer to make a choice with a lower expected outcome

LOSS 197

but a higher degree of certainty. We're prepared to settle for less if we have a firm belief that we'll gain something. A future, albeit bigger, pay day is less attractive in comparison.

When I think about prospect theory I think about that iconic (but unrelated) marshmallow experiment done at Stanford University in the early 1970s. Children were given the choice of one marshmallow now or two marshmallows if they were prepared to wait fifteen minutes. The temptation to go immediately for the one marshmallow was overwhelming.[4]

Psychologists interested in how humans respond to climate change have found prospect theory to be an extremely useful concept. In particular, it helps us make sense of why there's such a gap between belief and behaviour in response to climate change. We can agree climate change poses a future threat to our wellbeing and way of life. We can agree that the main way to cushion such a threat is to change the way we live and engage in various forms of 'sacrifice'. And yet on the whole we have been slow to do this.

Taking on board prospect theory, it makes sense that, as Marshall comments, 'people will be strongly disposed to avoid short-term falls in their living standards and to take their chances on the uncertain but potentially far higher costs that might come in the longer term'.[5] As we saw in the previous chapter, our optimism bias encourages us to put the best possible spin on the future. We're reluctant to do anything dramatic or pay anything significant now because even though we believe climate change will get worse in the future, we don't think it will affect us as badly as it will other people. Our brains make this high-stakes gamble seem like a safe bet.

It's not just that we feel this loss aversion at the personal level. At the level of big organisations and whole sectors of the economy, loss aversion is ever-present. There's no doubt that while a dramatic move away from fossil fuels will create new economic opportunities and jobs, there will be inevitable 'creative disruption', with many industries and companies detrimentally affected, some in the short term, some forever. The people leading those companies and industries might acknowledge, publicly and/or privately, the climate science, and commission reports from their corporate responsibility divisions on the topic. But they're often reluctant to make dramatic changes to their business practices in the short term.

Owners, executives and shareholders tend to regard the future in terms of quarterly profit and loss, and end-of-financial-year statements. That's not to say there hasn't been progress in the area of businesses acting on climate change and business leaders around the world speaking out. But the day-to-day, immediate agenda of so many in the corporate world is still driven by the pursuit of short-term profit and a conservative approach when rethinking business models in light of future climate effects.

And it's not just on the capital side that there's resistance to change. In a country like Australia, where the union movement still has significant power in one of our two major parties, the left-of-centre Australian Labor Party, the notion of job losses caused by such a climate-driven economic restructure is not a welcome one. The resistance from unions representing mining jobs has been significant. As a friend of mine in the union movement put it to me,

LOSS 199

'The union would rather have their members working in mining for the next ten years and stay in the union rather than leave the industry and the union and be employed for the rest of their lives.'

We cling to the status quo even though we know objectively that it will cause great harm, some of it avoidable, in the long run. Given this loss aversion and sensitivity to risk, it's not surprising that one of the few sectors that was first to highlight the risks of delaying action on climate change was the insurance industry. Its entire business model is based on what may or may not happen in the future, calculating risks based on data and projections over a human being's life span. Insurance companies around the world continue to warn governments, financial institutions and investors that failure to reduce emissions could result in a world that is 'pretty much uninsurable'.[6]

In thinking about loss as a barrier to action on climate change, we can cycle back to comments in previous chapters on issues such as political polarisation and the mindset of active and professional deniers. Namely, that climate change messages can feel like an attack on conservative values. For a conservative, to accept climate change as serious and caused by humans means a possible rethink or loss of their political identity. And rethinking your identity as a vocal climate denier can mean more than a loss of face. It could mean you lose friends, community, status, the opportunity to argue with family members at the dinner table, and (if you're a journalist for Fox News) your job.

As a result of her experience trying to convince an ardent climate denier of the veracity of the science (see Chapter 7),

Australian climate activist Anna Rose learned how important it is to 'acknowledge the loss that accepting climate science brings with it for those who hold a strong free-market ideology and are opposed to corporate regulation'.[7]

I would point out that there are plenty of heartfelt and genuine examples of well-known climate sceptics changing their minds and admitting their mistakes. There is a growing list of high-profile politicians, academics, meteorologists and journalists, all on the record as deniers or strong sceptics, who now accept the science and the need for urgent action (frustratingly, though, many of the politicians have done so only *after* leaving positions of power).[8]

If leading Republican consultant Frank Luntz (who famously advised his boss George W. Bush to switch the language from 'global warming' to 'climate change' so it would appear more natural and benign to voters) can publicly change his mind, other deniers can muster the gumption to do so as well. Then again, prospect theory tells us that loss of face today can be more upsetting than the loss of the earth in the future.

Finally, loss is ever-present in the language used to frame climate change for the public. Many climate communications experts argue that this relentless negativity, in and of itself, is a barrier to engagement for many people. Stoknes is critical of the way climate change is overwhelmingly framed in terms of loss.

> We've been told we'll lose beautiful forests, butterflies, birds and streams, and even human dwellings, coral reefs,

polar bears, snow and ice. Worse, the climate solutions have also been framed as losses to us. We're going to lose the possibility to travel where we want, eat meat, or shop freely. Environmental economists, too, have been prone to use the cost frame in many ways. The polluter should pay the true cost. A global tax should be put on emissions, raising the consumer costs.[9]

Talking about losses without also talking about gains means much of the 'good logic of the underlying economics' gets swallowed up.[10] We can't see that we gain much more than we lose by acting on climate change.

This is where events like the Australian bushfires of 2019–20 (or, for that matter, the COVID-19 pandemic) have the capacity to become, as Tony Leiserowitz calls them, 'teachable moments'. Calculating the costs of wildfires has one possible benefit. It makes the people who acknowledge some connection between climate change and extreme weather realise, really understand, there is a cost to inaction. It might encourage them to be more open to the idea of spending more taxpayer money on climate change adaptation and mitigation. It might push policymakers and lawmakers to be more confident in their arguments that this spending is necessary for public safety and economic security.

It allows us to recognise how difficult it is for human beings to cope with loss, how we prefer short-term certainties to what seem like long-term risks, how we want to avoid certain losses today rather than make sacrifices to avoid larger losses in the future. We can acknowledge that changing your mind on climate change can involve a loss of identity. I agree with Stoknes that in talking about climate change we have to find a way to discuss gains as much as losses. But as previous chapters on fear and despair have shown, turning a blind eye to the notion of loss doesn't work.

Coming to grips with climate change means understanding significant losses are inevitable. In fact, some are already happening today; just in my country, big parts of the Great Barrier Reef and the Torres Strait are suffering loss, but across the globe significant changes have been observed in Alaska and the Maldives, iconic cities like Venice and Rio de Janeiro, and regions like the European Alps and California's Napa Valley. As Australian researcher Susie Wang and her colleagues write, 'to care about climate change is paradoxically not about climate change itself, but about the things that it will harm or take away from us'.[11] In all the research I've done on people who care about climate change, part of their engagement with the issue has involved recognition of and reflection on what will be lost.

For me, climate change has taken away any number of my plans for the future. The fires destroyed many of the places I have taken my children on holidays. It seriously threatens to take away a safe and secure future for my children, impacting the prosperity and happiness of the community surrounding them. From time to time I find myself experiencing profound sadness and grief about these losses, albeit only silently or to a few friends going through the same feelings. Indeed, psychologists working with people who are suffering from eco-anxiety and depression, both in academia and in clinical practice, talk about the critical role of grieving not just over what is being lost now but what will be lost in the future.

My grief is nothing in the face of those people here in Australia and around the world who are grieving not just the loss of property, livelihoods and loved ones but the loss of 'home', the sense of place that has shaped their identity and culture. This is a manifestation of environmental grief called 'solastalgia', named and described by Professor Glenn Albrecht from Murdoch University in Western Australia.[12] It is a form of environmental grief generated by losing important aspects of one's home environment. You can see 'solastalgia' in the poetry of Lavetanalagi Seru from Fiji and the speeches of the people of the Torres Strait described in previous chapters. There is a growing body of research looking at the reactions of people whose homes have been forever transformed by climate change. In a sense, we get a preview of what's to come for the rest of us through understanding their responses to this loss.

One researcher who has examined this in detail is academic Susanne Moser, whose idea of hope as 'functional denial' I referred to in the last chapter. She conducted some focus groups with people living in coastal areas of the United States where sea level rises are claiming land, places where 'keeping what we've had' is not a realistic, long-term option. What she found through talking to these coastal residents is that some kind of tolerable outcome for the future was only going to be achieved with a focus on the common good. The only way forward for these communities was for them to tap into what they loved about the place they called home, to reaffirm their connections with each other as neighbours and to the non-human world around them, and to meaningfully engage as citizens in collective problem-solving.[13] She also found that in order to achieve this, people had to allow themselves and others to go through the range of difficult emotions—including grief and despair—triggered by the changing environment and loss of loved places.

Every religion has its own highly developed rituals to deal with grief and loss. In countries like my own, with a growing number of people who describe themselves as having no religion, we're more likely to go to a therapist than a church when we find ourselves trying to cope with these emotions. Sometimes that works, sometimes it doesn't. It can be a very introspective approach to coping with loss. What religion can provide to those who are grieving is not only rites of passage but a community of mourners. The collective nature of grief in religious communities can sometimes offer the kind of comfort therapy can't.

I've been interested in the role that religions play in climate change for some time. It sometimes comes as a surprise to some people I talk to that so many religions around the world have made strong, unequivocal statements about the urgent need for action on climate change. There have been Islamic, Hindu, Sikh and Baha'i declarations on climate change, and major statements from Jewish groups and rabbis across the globe. There is a Global Buddhist Climate Change Collective.

The Dalai Lama has described climate change as 'a problem which human beings created' and must be responsible for solving. But we are instead, he has said, 'relying on praying to God or to Buddha. Sometimes I feel this is very illogical.'[14] In 2015 the Pope penned an encyclical on ecology, which stated that the science of climate change is clear and the Catholic Church views climate change as a moral issue that must be addressed in order to protect the earth and its inhabitants.[15] In the United States even some evangelical churches have made strong statements on climate change; there has been an Evangelical Declaration on Global Warming and there is an active Evangelical Environmental Network.[16] What unites these different faiths is essentially the same argument Anna Oposa used to convince the Catholic Church in the Philippines to speak out about prayer balloons (see Chapter 6): that we are stewards of God's creation.

An emerging number of studies show that religious people engage with the climate issue in particular ways, shaped predominantly by their spiritual beliefs. For example, two large surveys by researchers from Yale found that Christian Americans say 'protecting God's creation' is the top reason for wanting to reduce global warming.[17] They saw the issue of environmental protection as a moral and religious one. Furthermore, their pro-environmental and climate change beliefs increased if they perceived that other Christians cared about environmental protection.

The conclusions from this study make me worry about the tendency of some in the left to either ignore the evident concern within faith communities about climate change, or attack the religious beliefs of climate deniers. This can be detrimental to the climate cause. Indeed, Finnish academic Panu Pihkala argues that religious communities have an important role to play in supporting people as they process their emotions and confront the existential questions triggered by climate change.[18] George Marshall agrees, arguing that religions 'embody long-term thinking, encouraging their members to accept responsibilities and invest in a legacy that extends far beyond their own lifetime on Earth'.[19]

But in my own social circles, when I suggest that faith communities might have an important role to play in the climate change cause, I've had more than a few lefties say to me, 'Aren't they the problem?' Australia's current prime minister is an evangelical Christian who belongs to a church that preaches the prosperity gospel and believes in the rapture. His religious beliefs are often the butt of jokes in signs carried around at climate protests. While there is certainly evidence that some religious denominations are more sceptical, even outright contemptuous, of the climate change science than others, that shouldn't lead us to ignore the religious consensus that exists.

Instead, we need to realise that religious leaders can play a crucial role as 'climate brokers', people who have an ability (because of their identity and place in society) to speak to different groups on climate change or appeal to certain communities who aren't already alarmed or active. Professor Andy Hoffman writes that religious leaders can be more effective spokespeople than non-religious leaders on environmental issues with faith-based communities. 'When people hear about the need to address climate change from their church, synagogue, mosque or temple, for example, they will connect the issue to their moral values.'[20]

And climate change communicators who are also religious can even inspire nonbelievers like me. One of my favourite speakers on climate change is a climate scientist and evangelical Christian Katharine Hayhoe. The daughter of missionaries and married to a pastor, she admits her life as a Christian and a scientist is 'like coming out of the closet'.[21] While she gets her fair share of trolls and attacks for her views, it's extremely hard to dismiss her as your typical communist, faithless leftie. She regularly speaks about the central role her faith plays in her climate advocacy. And you're on shaky ground debating the science with her given she has a PhD in atmospheric science. She doesn't fit neatly into the 'climate scientist' or 'pastor's wife' stereotypes, and is a more effective communicator because of this.

My interest in the nexus between religion, faith communities and climate led me directly to the Reverend Jim Antal. It was in talking to him that I realised people of faith have a lot to teach a lapsed Catholic girl like me about how to manage loss and grief over the fate of the earth. Reverend Antal is a leader in the United Church of Christ (UCC), a climate activist, author and public theologian. The UCC was founded in Massachusetts in 1620 by the pilgrims who fled religious persecution in Europe, and has been one of the most progressive churches in America—the first to ordain women, people of colour and openly gay men and women.

An environmental activist from the first Earth Day in 1970, Reverend Antal has been outspoken throughout his long career on environmental issues of all kinds. He has protested outside the White House, one time even cuffing his hands to the fence. He was arrested in Boston as he and other activists sought to block the construction of a new natural gas pipeline. In July 2013 Antal wrote and championed the UCC's resolution to divest from fossil fuel companies, the first of its kind in the country. Throughout his time as a leader in the church, he has encouraged hundreds of other pastors to preach to their congregations about climate change.

In his book *Climate Church, Climate World: How people of faith must work for change,* Reverend Antal contends that we need to reimagine the role of church communities and their capacity to confront and resolve the problem of climate change. He argues persuasively that Christians can't continue to emphasise personal salvation while ignoring collective salvation. 'If we continue to reduce the Creator to an anthropocentric projection who privileges and protects humanity ... then the practice of religion will continue to diminish and it will add little to the redemption of creation.'[22]

While I was visiting Yale, I contacted Reverend Antal through his website and to my delight was quickly granted an interview. He gave me a generous chunk of time over the phone from his home in Massachusetts. He grew up in a family of scientists, and his early years of social activism were focused on the campaign to stop nuclear weapons. He first learned about the greenhouse effect from reading about it in *The New York Times*, and then preached his first sermon on climate change in 1988, around the time of the James Hansen testimony, when writers like Bill McKibben and others were publishing their first books on global warming and its consequences. He now recalls that the reaction to this first sermon from his congregants was mostly bewilderment.

> Everybody was aware of environmental concerns like water and air pollution but I think at that time there was very little awareness that the entire project of creation was at stake. People didn't push back because it was a very progressive town, but people were just aghast.

Despite decades of activism in a church that has been speaking out on environmental issues since the 1950s, Reverend Antal still finds it challenging to convince people that faith groups can have a progressive position on climate change. For example, when he was trying to find a publisher for his book, the general response was that religious people wouldn't be interested in climate change. 'They said to me, "Evangelicals are on the side of the science-denying, conservative ideologues, why would we publish this book?"'

In June 2015, when President Trump pulled the United States out of the Paris Agreement on climate mitigation, Reverend Antal authored a resolution declaring a new moral era in opposition to this withdrawal. The national UCC Synod, the church's peak decision-making body, passed that resolution with a 97 per cent supermajority. It made headlines across the country. Suddenly there was evidence that Christians cared about the climate, and publishers came knocking on the reverend's door.

It's still the case that people are a little surprised when I tell them I'm a minister and my work is in faith and climate change. They go, 'Really?' And typically what I say to them is, 'How about Psalm 22 Verse 1: "The earth is the Lord's, and everything in it, the world, and all who live in it"?' And then people begin to go, 'Oh, if it's God's earth and we're wrecking it, I guess that's a religious issue . . . Oh, now I get it.'

Reverend Antal's reflections on loss and grief have stayed with me, making me recognise the enormous value of having faith groups involved in the climate change cause:

I think people of faith are more willing to talk openly about grief, and ministers are willing to preach about grief as they preside over funerals. When people gather for worship every week, there is always something that is grievous that is offered up for prayer. Faith communities are communities

LOSS 211

in which grief is regularly talked about. It's not that secular people don't grieve, they do, but specifically grief for secular people is something that may isolate them, something they need to process themselves, or perhaps with their partner. Whereas faith communities, one of the reasons they exist is to help people process their grief with others.

Reverend Antal is a mentor and a guide for so many other pastors looking to preach about climate change to their congregation (which is significant given UCC members amount to about 1 million Americans). In 2006, when he was the head of the church in his state and responsible for leading 900 clergy, he would explain to them that increasingly, they would have to return to climate change again and again in their work.

> I gathered ministers in a group, asked them to line themselves up and to walk to a certain side of the room if they had preached about climate change at least four times in the past year, stand in the middle if they had preached on it once and over the other side if they had never preached on it. And of course, in 2006, everyone went to the 'never' side. But then I would say to them, 'Listen, you all need to understand something: if you don't begin to preach on climate change at least once a month, I don't know if it will be ten years from now or fifteen years from now, every single service will be on grief and the only reason people come to church will be to grieve the planet.'

He believes that if churches are going to be relevant and useful in the coming years, they will have to go beyond what he calls the 'superficial continuations of what the church has been for decades'. 'Everything has changed and that requires a shift in the fundamental understanding of the vocation of the church,' he explains. 'We have to shift it in a way that mobilises people of faith to act on behalf of God's creed.'

Reverend Antal believes that faith communities and environmentalists have much in common in terms of their concern about injustice, and that they can become firm allies in the climate cause. 'Every Wednesday night we are feeding the homeless, the hungry, every Saturday night housing the homeless. Let's talk about how climate change is going to make instead of 1 billion hungry people on the planet, 5 billion hungry people on the planet, instead of having a migration of 3 or 4 million people from Syria, there's going to be migrations of 2 or 3 billion people from the southern hemisphere.'

Finally, I asked Reverend Antal his thoughts about hope. Of course, when tragedies occur, either natural or not, we always hear faith leaders respond with 'hopes and prayers'. It turns out he defines hope in much the same way as others I've mentioned in this book—as a kind of 'engaged activism'.

> I'm really explicit about that fact that hope has nothing to do with optimism. We need to look at the worst possible scenarios, right in the face. But we also need to be filled by the beauty and wonders that creation reveals to you

every day. I think to some extent that's the compensation for the hard and sometimes depressing work on climate change. You're banging your head against the window and then you look up and you see the sun rises in the morning. You think to yourself, 'Oh my God, we get to live on this planet? Let's get busy.'

George Marshall argues in *Don't Even Think About It* that the 'Green Team' can learn a lot from the 'God Squad' when it comes to emotional and ethical appeals to the community to act on climate change. The importance of sharing emotions (particularly negative ones associated with fear, grief and doubt) with a community of believers. The importance of creating a legacy for future generations, thinking long-term and valuing things beyond the financial and superficial. How to show generosity, kindness, forgiveness, compassion to others.[23] Looking at how successful faith communities function, those of us in the climate movement 'could learn to find ways to address feelings of blame and guilt that lead people to ignore or deny the issue, by enabling people to admit to their failings, to be forgiven, and to aim higher'.[24]

Of course, religious communities are not unique in their ability to achieve these goals, but they've been at it, many of them, for hundreds of years. They have a few lessons to teach the faithless. While there has been a collapse of trust in religion generally for all kinds of (legitimate) reasons in countries like Australia, that doesn't mean we can't learn from people like Reverend Antal, even if we don't share the same commitment to God.

I'm not a religious person but I understand the value and allure of a religious life. I deeply respect faith leaders like Reverend Antal who are using their leadership role to preach on climate and try to find ways to evolve the mission of the church to address the very modern challenge of climate change. I also understand the need for rituals, both secular and religious, to mark important moments in our lives, to allow us to process our emotions and to share that with others. In addition to my day job, I'm a registered marriage celebrant, which means I can legally conduct non-religious marriage ceremonies as well as other non-legal rituals such as funerals.

At a celebrant training session I did recently, I found out about the increasingly popular eco-burial movement. This involves preparation of the dead in environmentally friendly materials or cremating the body and, instead of placing the ashes in a traditional plot or tomb, burying them in a dedicated natural space. For example, in California's Mendocino County, 8 hectares (20 acres) of redwoods, firs and oaks called Better Place Forests functions as both a memorial and a carbon sink.[25] The dead are cremated, and the bacteria-free bone dust mixed with dirt and placed at a tree's root system in a 'spreading ceremony'. By purchasing a tree instead of a burial plot, you invest in the preservation of the landscape. Each purchase also triggers an 'impact trees' program, which commits to replanting a certain number of trees—between 25 and 400 depending on the price of the purchased tree—in wildfire-affected areas of California.

In her book *Greening Death*, Suzanne Kelly argues that these kinds of eco-burials can provide a new way for people to cope with climate anxiety.[26] I know the choices I've made in my lifetime have brought about environmental damage (I've tried but I know I have not always walked softly upon this earth). The least I can do is shuffle off this mortal coil without the toxic chemicals and waste of a conventional funeral and let my ashes feed the roots of a tree turning CO_2 into oxygen.

I like the idea of being turned into ash, folded into rich earth and buried in a carbon sink. Perhaps my death, the loss my children will feel about that, will be softened as they stroll through the forest and inhale the scent of eucalyptus or pine. All the more reason for me, while I'm alive, to do whatever I can to ensure trees will grow and remain standing on a planet altered by climate change.

CHAPTER 11

LOVE
Or do it for the birds

There are not a lot of psychological studies on how love influences our beliefs and behaviour regarding climate change. But love hovers in the background of many of the studies I have already described in this book. If you fear loss, it's the loss of something you love or care about. We feel guilt and shame when we let the people we love down or disappoint those whose good opinion we value. Any threat to the people or things we care about can provoke feelings of anger and fear. Our fight or flight system is there not just to protect ourselves, but our children and our homes as well.

All the people I've interviewed in this book have come to climate activism because climate change threatens to destroy something they love. For me, doing something about climate change became my life's purpose because I love my children and I want them to have a happy and secure future. I believe climate change threatens that future. I've made the connection between climate change and the things I love. But as with everything to do with communicating about climate change, making this connection is not easy. Just asking people what they love and then saying, 'Climate change will harm it,' is rarely enough to convince them. As teen activist Daisy Jeffrey said in an early chapter of this book, coalminers love their children too. So, I assume, does Rupert Murdoch.

What love has going for it is that, in the realm of emotions, it's as close to unequivocally positive as you can get. By love, I don't mean romantic love (*pfft!*). I mean deep and abiding affection, steady attachment, and unshakable commitment to a person, place or thing. Those writers and researchers looking at effective climate communications stress the importance of 'positive frames' in trying to overcome some of the cognitive biases and psychological barriers in the way of belief and action on climate change.

Per Espen Stoknes has written extensively about the power of positive strategies to help avoid triggering the 'emotional need for denial through fear, guilt and self-protection'.[1] He writes that our conversation on climate change must be engaging and inspiring, and bolster feelings of community. He argues that appeals to people to care and act on climate change should tap into positive human emotions like the love of beauty and pleasure. He also argues that we need to frame solutions to climate change in similarly positive ways: it 'works so much better when people want it, like it, love it rather than when they implement it by duty, guilt, rule, or fear of punishment'.[2] Emphasising this connection between acting on the climate and protecting what we love helps us show that we will gain as much as we will lose, as I explored in the previous chapter.

Writer and Nobel laureate Elie Wiesel once wrote that the opposite of love is not hate but indifference. Love is when you really, truly care. Care is such a small yet powerful concept. It's a fundamentally active word. It means you will do what's necessary for the health, welfare, maintenance and protection of someone or something you value. In some ways, it's easy to be a lover. It's a tough, tough job being a carer.

One of my favourite psychological studies on care and climate was conducted by a team of Australian researchers including former premier of Western Australia and federal government minister Carmen Lawrence (now a university professor). The researchers wanted to see if the strong emotions triggered by a perceived threat to what participants personally cared about could motivate them to care about climate change itself, which would in turn predict climate policy support.

When we say we care about climate change, that's actually a misnomer. We care about what climate change will do to the things we care about. The group of Western Australian researchers developed the term 'objects of care', an incredibly useful concept because it can refer to pretty much anything that matters to people. It needn't even be an object.[3] They argue that objects of care have the capacity to be 'connectors', making the issue of climate change seem personally relevant to an individual, undermining the distancing effect that makes us feel climate change is far away in time and place.

The researchers took the unique approach of comparing the emotional responses of climate scientists to the question 'How do you feel about climate change?' with those of university students and of the general population. I started my chapter on despair with a description of how climate scientists are coming out about their climate-induced depression and anguish. The researchers from Western Australia reported similar findings.

The 44 scientists who responded saw strong connections between acting on climate and the welfare of future generations, and between human beings and the natural world. They saw their relationship with the planet as akin to a longstanding and deep friendship, and reflected on the pain they felt seeing that this 'friend' was seriously ill. 'Imagine how a medical doctor feels having to inform their patient, an old, life-long friend, of a dire but treatable diagnosis,' one scientist wrote. 'There is a similar closeness between climate scientists and the planet.'[4]

In their responses, they covered the whole spectrum of human emotions and spoke from different points of view—as scientists, citizens, humans, parents, grandparents and climate messengers. The scientists were steeped in an understanding of the scientific facts, but instead of responding by reiterating the data, they felt free to explore their complex emotional response to climate change. At the heart of it all was the love they felt for the planet and the people living on it. You didn't need to convince them of the urgent need for effective climate policy.

The university students and general citizens mentioned a number of objects of care in their responses—'future generations' was the most common, but specific places also came up, such as the Great Barrier Reef. Interestingly, the emotional response from the students and general population was more restrained than that of the scientists. Those in the general sample who expressed strong negative emotions in their responses, such as anger, shame and fear, were significantly more likely to support climate change policies. This group also mentioned objects of care with the greatest frequency. For them, climate change was emotional, personal, relevant and urgent. They could see it was going to impact places and people they cared about.

The researchers concluded:

'Objects of care' that link people to climate change may be crucial to understanding why some people feel more strongly about the issue than others. These 'objects of care' may bridge the psychological distance between the self and climate change, making the issue of climate change seem more personally relevant, evoking stronger emotions, and prompting action.[5]

In my own research, we often ask focus group participants to rank lists of issues in order of importance. I've noticed that they put climate change in the top three if they see a direct connection between it and other issues of importance to them. For example, if they're concerned about health and have already come to the firm conclusion that climate change exacerbates health issues, then both health and climate change get put at the top of their list. Often that high concern about health is driven by personal circumstance, like having a child with asthma. In contrast, if they see climate change as disconnected from whatever issue matters to them—say, the economy or national security or transport—then it gets put at the bottom of the list. Those people often say to me, 'Sure, I care about the environment, but we have to put people first.'

What these studies show is that the starting point in any effective discussion of climate change is not climate change itself, but what we care about, what we love. That is, of course, different for different people. But, as the song says, everybody loves somebody (or something) sometimes. One of the terrifying things about climate change is we know it will alter everything. One of the upsides of that is that we can find a way to connect it with anything that matters to anyone.

Some climate activists see this focus on starting the conversation with an individual's object of care as a narcissistic or narrow approach to the climate change issue. Climate change only matters to you because it's going to impact *your* kids or *your* house or *your* job. But the object of care is merely a starting point, a way to make climate change personally relevant to people struggling to understand what it means for them. And we should not forget that love, true love, is not selfish. Indeed, one of the many synonyms of love is altruism.

Our definition of love and care in this context necessarily includes our attachment to our group, our tribe, our people, our community. George Marshall writes that despite our massive cultural and social differences, there are certain things we all share as human beings, and one of the most powerful is our 'deeper instincts to defend our family and tribe'.[6] Studies show that our love and care for our tribe is more persuasive and more motivating than mere self-interest.

One fascinating American study shows that the impact of climate change on our community can be more influential than the impact it has on us personally. Researchers from the University of Colorado Denver and Duke University found that while people were more likely to believe in global warming if they lived through the Colorado floods of 2013, the level of personal damage had no statistically significant impact. In other words, having their house destroyed didn't shift their beliefs more than those of someone whose house merely got damaged. What mattered most was the shared experience of the community, what happened to their friends and neighbours, rather than just what happened to them.[7]

This civic pride, community spirit, whatever you might call it, is what makes us pull together in a crisis, donate money to people we don't know, and demand governments and corporations do something on their behalf. Love can expand beyond the circle of our children, family and friends to encompass a community, a society, a city, a movement for change. Encouraging and sustaining this notion of love creates greater momentum around climate action. And it will help us cope with what is to come.

As I said, objects of care can be anything: people, things, places, practices, professions, hobbies and obsessions. The northern cardinal, or *Cardinalis cardinalis*, the official bird of seven US states including Ohio and Kentucky, can be an object of care. And the love of birds can make you care about climate change. At least that's what Lynsy Smithson-Stanley taught me.

I met Lynsy when I was visiting Yale University and she was working at the Yale Program on Climate Change Communication on their media and education strategies (she has degrees in mass communication and journalism). But it was her work with birds that intrigued me the most. The program's website at the time said she had done work 'activating bird-lovers around the climate threat', which 'included developing a research program to engage conservatives'.[8] Birds? Conservatives? It seemed like a pretty niche topic even for academia. I had visions of an army of men and women birdwatchers with binoculars in one hand and megaphones in the other. 'There are no hooded warblers on a warming planet,' they would yell. (I jest, but a quick google shows that a staggering two-thirds of North American birds are at risk of extinction due to temperature increases.)

While Lynsy's work stood out against the other Yale projects looking at big data, politics and public opinion polling, I had a gut feeling it might reveal some useful insights about how to make climate change tangible for people. Little did I know that Lynsy and her birds would teach me more than one incredibly valuable lesson about climate communication.

Lynsy is a bright and bubbly woman with girl-scout levels of enthusiasm and generosity. She became engaged with the climate change issue when the cap-and-trade legislation aimed at reducing carbon emissions was rejected by the US Senate in 2010. 'We didn't just shit the bed politically,' she told me frankly. 'We shit the bed communications wise.' This epic fail encouraged her to focus her career as a journalism and communications expert on finding better ways to talk to people about climate change. Instead of working with one of the mainstream green groups, she gravitated towards a membership organisation, with the idea that grassroots advocacy on climate would be an interesting challenge. She found herself in the role of director of climate communications and strategy at the National Audubon Society, which works to protect birdlife.

The Audubon Society is over 100 years old, making it one of America's oldest organisations of its kind. Named after a famous bird illustrator, John James Audubon, the first chapter of the Audubon Society was created in 1885 by a group of ticked-off fashion lovers. Boston socialites Harriet Hemenway and Minna B. Hall were angered by the impact on local bird populations of the trade in feathers for use in hats and other clothing. The cousins organised social events with other wealthy local women, encouraging them to boycott the garments in order to protect birds like the egret. Their afternoon-tea activism worked, and other chapters of the society spread throughout America.

When Lynsy joined Audubon, she was looking for ways to engage the membership on the issue of climate change. It wasn't a particularly easy sell. Audubon members come from across the political spectrum, liberal to conservative to independent to 'Go away, it's none of your business'. 'Sometimes when I'd talk to members about climate change, they'd tell me, "I go birding to get away from politics." They weren't interested.' Convincing the leadership of the society also required some effort:

> The board had been concerned about talking about climate change. The usual environmental NGOs in DC, their membership is all progressive. Our membership lists are different, more bipartisan. You can't assume they all agree on climate change or are ready to become activists.

engage the membership, Lynsy devised a research campaign structure aimed at understanding what with conservative members in particular. The to commission some bird-specific climate change n the political polarisation around climate change on both the science and the scientists, Lynsy and felt that bird science commissioned by the society itself would be more trusted by its members.

The result was a groundbreaking study, released in 2014, detailing the impacts of climate change on 588 North American bird species.[9] They created maps showing where different species were threatened across different parts of the country. The breadth and detail of the study allowed members to focus on the birds and places they loved to see them, specifically local birds that meant the most to them.

The second step was to commission a research agency to look specifically at the views of Audubon's conservative-leaning members on climate change and its impact on birds. They found the trust and authority of the society meant it was in a unique position to talk to conservative-leaning members about climate change, especially those who were not currently engaged with the issue. They also found that while these members weren't sure of the degree to which humans were causing climate change, they still agreed with the need to act responsibly, especially when it came to the welfare of birds. 'Birds are the catalyst for action,' the researchers concluded. 'Audubon members are passionate about birds and are able to view environmental policy in a manner that cuts through partisan politics, largely due to their passion for birds and the outdoors.'[10]

The research showed that positive, hopeful messages were best in talking to these members about climate and birds, along with appealing to their identification with their local area and state (as Virginians, say, rather than just Americans). Subtle religious messages were effective as well, pulling at members' heartstrings by telling them that birds are in trouble and we have a moral duty to act (as stewards of God's creation). The research also identified the aspects of the climate debate that turned these conservative members off—namely, what they saw as dramatic and extreme predictions about the future.

Fundamentally, the study showed that if you kept focusing on what was good for the birds, you could get the members to support a range of pro-mitigation policies, such as limiting carbon pollution from power plants, and setting a federal price on carbon. Extraordinary, given these conservative members generally vote for a Republican Party that has, on the whole, resisted these kinds of policy ideas at the state and federal level.

The positive reaction of the society's more conservative membership encouraged the society to go ahead with extensive training and advocacy around birds and climate change. They invited members to come to training sessions, to take them through all the research and give them guidance on how to talk about climate change and encourage others to act. To get political about birds. In these sessions, Lynsy saw that connecting birds with climate change could bring up some pretty strong emotions among the membership:

> We'd talk to them about the bird science we commissioned, trying to connect their favourite birds with climate change. And you'd get these instant reactions. Like, 'I've always seen the scarlet tanager in this part of Ohio but these maps show me that soon they won't be around in the summer.' And that deeply affected people. I saw grown men cry.

The training sessions were very successful, with society members becoming politically active at every level, doing everything from sharing social media posts to directly contacting politicians. They developed a quick cheat sheet for Audubon members to talk to other bird lovers about climate. The first step? Make it about birds:

> Climate change can seem like an overwhelmingly complex and abstract issue. But we don't need to tell the whole

story of climate change with elaborate temperature graphs or atmospheric data. Tell people that climate change affects beloved species, like the bald eagle, wood thrush, or rufous hummingbird, and tap into their love for these birds.[11]

It also suggested they make it local, personal, hopeful, and avoid extreme language. Finally, it advised not getting bogged down in scientific details or in arguments about the exact causes of warming. What mattered was that their beloved birds were being threatened and so they needed to act.

Lynsy's work with conservative bird lovers taught me a valuable lesson about how we can talk about climate change with other people. Getting too bogged down in the science, requiring people to share all of your values or beliefs, is unlikely to get them to act. If you can find something that matters to them and then connect that meaningfully with climate change, that's a critical first step. If you can get them to act in concert with other people who have the same passions and devotion, that's even better.

If you can, try to avoid or sidestep some of the aspects of the climate discussion they dislike that would shut the conversation down (too much doom and gloom, dramatic predictions about the future), then you can keep the conversation going. And always, always keep that object of care at the centre of it all. 'People could disagree with the extent of human involvement in climate change but they could agree on the solution because we connected it to the protection of birds,' Lynsy told me. 'Birds were the gateway drug to get them interested in climate change.'

Being a marriage celebrant, I'm more than familiar with the famous passage from 1 Corinthians 13, a favourite reading at weddings, both secular and religious:

> Love is patient, love is kind.
> It does not envy, it does not boast, it is not proud.
> It does not dishonour others, it is not self-seeking, it is not easily angered, it keeps no record of wrongs.
> Love does not delight in evil but rejoices with the truth.
> It always protects, always trusts, always hopes, always perseveres.

It's never been my favourite reading, but I've taken to revisiting it in the context of my thinking about climate change. I can see articulated in it the kind of love we need to develop in the climate era, and so it's taken on new meaning for me. The value of love that is as much about others as it is about yourself; the desire to protect and the need to persevere; not becoming easily angered and keeping resentment to a minimum; the value of a steadfast hope even in the face of trials and tribulations. Rejoicing in truth.

Love as the starting point and the final destination.

CONCLUSION

TALK ABOUT CLIMATE CHANGE
It's the right time

Climate change must be one of the hardest topics to talk about, surpassing sex and drugs and religion, on a par with topics such as death and depression. I have to admit that even I, despite having spent a solid two years thinking about how to talk about climate change, find it hard to talk about it with the people I love. I don't talk about it with my five-year-old twins, of course, although we do talk about recycling and how beautiful trees are, and we read *The Lorax* together. I do talk about it from time to time with my eldest daughter, who's now eleven years old. 'I know, Mum,' she tells me, somewhat exasperated when I mention it. She's learned about it at school and it's discussed among her friends, but there are no signs that she's feeling any of the eco-anxiety other young people are experiencing around the world, which pleases me.

I find talking about it with extended family members difficult. I'm worried I'll come across as judging their lifestyle choices, introducing tension into already complicated relationships.

Sometimes I wonder if the conversations I have with friends about it are making me a less than attractive dinner companion. Once upon a time we would get together to eat, drink and gossip about work and mutual acquaintances. Now I find myself telling them about melting permafrosts in Alaska.

I know I'm not the only one who thinks about climate change all the time but finds it hard to talk about in daily life (I can talk about it endlessly in a professional capacity, believe me). George Marshall writes that surveys coming out of places like the Yale Program show that the vast majority of us rarely or never talk about climate change, even inside our close circle of friends and family members.

> Women talk about it far less than men do, and as a group, younger women talk about it less than anyone, especially . . . those with children. Another survey found that a quarter of people have never discussed climate change with anyone at all. In real life, it seems that the most influential climate narrative of all may be the non-narrative of collective silence.[1]

In my own research work I've started asking participants in surveys and focus groups if they ever talk about climate change and, if they do, why. On the whole the answer is no or rarely, and only when the weather acts like a prompt, for example if it's an unseasonably hot day in spring. When presented with a survey question by a pollster about whether we're concerned or not about climate change, the majority of us might tick the box that says yes, but we won't be anywhere near as enthusiastic about bringing the topic up in polite conversation.

And no wonder. It's probably one of the most, if not *the* most, difficult topic to talk about, because it's so political, emotional, intense and overwhelming. It seems to require expert knowledge about everything from weather patterns to physics to economics and history. It can be even harder to talk about with people you work with or are just friendly with. You might have a nice and easygoing relationship with a neighbour or one of the mothers at the kids' school or the guy you have to talk to in Accounts, then one day you mention climate change and suddenly you're in a tense discussion or they look at you blankly like you're mad or, even worse, you find yourself in an argument. That easygoing relationship has changed and suddenly you don't make eye contact with them anymore.

You might think that not chatting about climate change at the gym or barbecue get-together isn't such a big deal. But breaking the climate silence is critically important to helping circumvent some of the cognitive biases and psychological barriers we've examined in this book. For example, in the chapter on fear, I wrote about our tendency to focus more on risks that are being talked about in our social circles than on those that are less talked about, even if the ones we don't discuss are more objectively credible.

So the silence around climate change in our daily lives works to decrease our sense that it is a real and imminent threat, which in turn affects everything from our voting decisions to our purchasing choices and general behaviour. This goes for the silence in our daily lives as well as the lack of consistent attention to it in broader conversations at the community and company level and in the mass media.

Furthermore, think back to the chapter on denial. For those people engaged in passive denial, as Stoknes points out, 'the issue [of climate change] becomes unspeakable [and denial] offers itself as a convenient way out of the discomfort'.[2] Silence is both a symptom of this passive denial and allows it to continue unchallenged. On the flip side, we have also seen how conversations between family and friends can be extremely effective at persuading people of the importance of the climate issue, even cutting through political partisanism. One research study I mentioned shows that simply having people around you who care about climate change can make you more likely to care about it.[3]

This is why researchers looking at psychology and climate change stress the importance of talking about the issue with other people. Climate scientist Katharine Hayhoe dedicated her TED talk to exactly this topic, calling it 'The most important thing you can do to fight climate change: talk about it'. The climate scientist, mother and evangelical Christian encourages us not to worry that we don't have all the climate science at our fingertips when we speak up on the topic. 'We don't need to be talking about more science; we've been talking about the science for over 150 years,' she rightly points out.[4] Instead, she encourages us to start our conversation with others from the point of view of shared values and, to use the terminology I wrote about in the previous chapter on love, 'objects of care'. Our community. Our concern for the economy and jobs. Local green spaces. Animals (remember the birds from the previous chapter).

Hayhoe's starting point for her own conversations is often her faith, but it's also the love she has for her kids, her role as a mother. What if you don't know what matters to someone you want to talk to about climate change? Well, ask them, Hayhoe suggests. First have a conversation about what makes them tick and then connect the dots between 'the values they already have and why they would care about a changing climate'.

I promised at the beginning of this book that reading it would help you better understand how the people around you are reacting to climate change. And that you'd also get a better understanding of your own complex and shifting reactions to the issue. These two together—understanding your reactions and how other people react—should equip you with some of the insights and skills you need to start breaking this climate silence in the world around you. I'm not saying this is easy, but it is necessary.

There are lots of materials out there in internet land that can provide you with handy hints about having productive climate conversations. Both Stoknes and Marshall end their books with suggestions as well. Common to all the advice are some of the following principles I've touched on throughout the book:

- Focus on local issues.
- Make it personal and relevant.

- Try to find mutual points of empathy and connection.
- Try to highlight what we have to gain from action and not just what we'll lose.
- Try not to focus on blame or villains, or set up an 'us versus them' dynamic.
- Stress cooperation and unity, groups of people working together to bring about change.
- Focus on the present rather than just the future.
- Be honest about the size of the threat but try to avoid too much extreme, catastrophic language (a delicate balance there, I admit).
- Recognise people's negative feelings of anger, grief, anxiety and guilt.
- Understand that this is hard, no matter who you are and what you believe.
- And encourage a form of active hope, despite what the science tells us.

In his final comments in his book, George Marshall gives us a laundry list of ideas about how to communicate effectively about climate change, but one edict jumps out at me as particularly important: 'Never assume that what works for you will work for others.'[5] There are a billion pathways towards an understanding that climate change is real and requires an immediate and adequate human response. What works for me will not work for other people, even if we have a lot of other things in common. Climate change is a complex, all-encompassing and fast-moving phenomenon. As such, as David Wallace-Wells points out, 'it is not a subject that can sustain only one narrative, one perspective, one metaphor, one mood'.[6] Or to hark back to what Tony Leiserowitz said in one of the early chapters of this book, not everyone has to walk the same path that you did.

The process of writing this book has helped me develop my own set of guiding principles, covering a combination of helpful ideas on how to talk to others and personal rules for living in the climate age. They might help you too.

1 Listen and understand

I've spent a lot of time in this book showing how attitudes to climate change are not a reflection of people's understanding of the climate science but of their worldview, values and identity. And so it makes sense that before you even have a conversation about climate change with someone, you need to understand who they are, not just what their views on climate change are. Change can happen through conversation and debate, but first we need openness, empathy and a determination to find some common ground. Otherwise, we're just shouting past each other.

2 Talk about it

See above. Talking about it is hard, but the more you do it, the easier it becomes. You might actually find that with some people, it's less stressful than you imagined, that they might be feeling the same way you do. Find already existing groups or gatherings in your own life where there's trust and understanding, and see if it's possible to find a way to start the conversation—for example, suggesting a climate-themed novel for a book group. How about suggesting a screening of the movie *2040* to your local parent and teacher organisation? Or if you're in an outdoor sports club that's having its schedule interrupted by extreme weather, making an appointment to see your political representative and raising the issue with them? Be inspired by Tennis Australia, the sport's national peak body, which has spoken out about climate change loud and clear.[7] Start a conversation about climate in your professional networks and associations; in Australia and around the world we have seen numerous professional groups created around climate action—engineers, architects, farmers, lawyers, accountants, teachers, veterinarians and a range of medical professional groups speaking out. Just in this book we've seen how Lynsy Smithson-Stanley helped turn birdwatchers into climate activists, and how Cassia Read brought together local parents who loved to craft so they could talk about climate.

Marshall argues that in talking about climate change with others you should always try to keep an open mind, be alert to your own biases and be prepared to learn from other people.[8] Know that this might be one in a number of conversations you could have over time, without any guarantees of success (in fact, don't focus too much at the start on what success looks like). But just consider Daisy Jeffrey, whose conversation with a father and his two sons didn't convince the father but encouraged the two sons to join the climate strike. And Anna Rose, who was never going to change

Nick Minchin's mind but reached a vast and more open-minded audience with her passionate beliefs about climate change.

3 Start with love

You don't need to start a conversation about climate change with climate change. You certainly don't need to start it with a recitation of all the scientific data. Start with an understanding of that person's 'objects of care' (see Chapter 11), something you either already know about or will find out through talking to them about their values and interests. Marshall puts this another way, by suggesting we 'relate solutions to climate change to the sources of happiness, and the connections we feel with our friends, neighbors and colleagues'.[9] This reinforces that we have more to gain than lose from doing something about climate change: the protection of the people, places and practices that matter to us more. And the focus on love and care is a powerful counterbalance to the usual 'doom and gloom' associated with so much information on climate change in the media, and the toxic conflict in so much political 'debate' on climate as well.

You can extend the idea of 'start with what you love' to your own actions on climate. When people say to me, 'I'm worried about climate change but what can I do beyond renewable energy in my house and riding a bike to work?', I always ask them, 'What are you already doing? What activities and places do you care about, engage with on a regular basis and love? Well, start there.' You don't have to join an environmental NGO to be a climate activist (although feel free to do that too). You can express your concern for the climate through your life as you live it today. Look around your street, neighbourhood, workplace, sports club, church, social group and community organisation.

Jim Antal led the charge for his church to divest its investments away from fossil fuels. A dedicated dad helped organise solar panels for our local school (with a bit of help from a government grant). And whatever you decide to do, communicate that to your local politicians. Show them that it's not only groups like Greenpeace that care about the climate.

4 Work out what you can do without

While I'm not yet into Deep Adaptation and preparing for imminent social collapse before my children leave school, I do find a lot of questions Jem Bendell asks (see Chapter 8) extremely useful. 'What do we need to let go of in order to not make matters worse? What can we bring back to help us with the coming difficulties and tragedies?'[10]

Reflecting on these questions has helped me sort out my priorities in life. For example, it's more important to me that I travel within Australia, spending money in communities devastated by wildfires, keeping my carbon emissions low and developing in my children a love and respect for the Australian environment. I will choose those holidays over travelling to Bali or Europe or America. Let's face it, that's a very privileged form of sacrifice, but it's just one in a series of decisions I will make about how I spend my time and money. And choosing to let go of things you previously thought would be part of your future is going to be good practice, preparation for a time when we'll be forced to adapt regardless. This is all about preparing for the worst while still hoping for the best.

5 Hope is other people

The existentialist Jean-Paul Sartre famously wrote in one of his plays, 'Hell is other people.' But he was a moody bastard who was no fun to be with at parties. I've reworked his cynical phrase as 'Hope is other people'. All the research confirms the power of groups, small or large, to bring about change. We have seen in my interviews with activists like Chido Govera and Cassia Read, as well as the psychological studies, that people, including climate activists, develop their hope out of working together with others. Indeed, Stoknes concludes that sceptical activism is the best kind of hope we can develop for the climate age, and that activism is most effective if pursued with like-minded people. Using the power of social networks, however you define them and wherever you find them, is critical in building a social consensus on climate action and making that consensus visible to others.

6 Get to know your neighbours

Academic Susanne Moser expressed this sentiment, in her interview with *Dumbo Feather* magazine, in response to a question about the importance of community:

> There is no doubt that the harsh conditions we're currently creating will make us dependent on each other in ways we

don't even know yet. We're so focused on, 'Can I protect myself from this? Where can I move?' As if there is a place to hide from this global change. But to have any chance of surviving as a species, we need to share resources . . . and we're going to get a lesson in dependence and interdependence like none of us have seen.[11]

While COVID-19 has taught us much since these words were spoken, we should still think of our pandemic experience as just a dress rehearsal.

Moser's words make me realise not only the importance of investing and reinvesting in groups around me that sustain my family and my community, but also something perhaps more profound about hope and mobility. Like every Australian who isn't First Nations, my family came here as migrants, inspired to travel around the world to a different continent, with aspirations for a better life. They found that life, built it with their bare hands, with each generation becoming more financially secure and better educated than the previous one. And it's very common too that ambitious people leave Australia for big cities in America, Europe and Asia to seek fame and fortune.

Climate change threatens to reverse that process. This is already happening, with wars and civil conflict exacerbated by climate change creating a massive global movement of refugees, and rising sea levels and extreme weather events leading to increasing numbers of climate refugees. But any thoughts I might have about where I could go to 'escape' climate change, in the hope that I can protect my family, are of course naive. Some places will be less habitable than others over time, but no place will be 'safe'. The world is too interconnected and climate change too all-encompassing.

Only billionaires can indulge fantasies about living in space or in biodomes on heavily guarded islands. (Personally, I would rather perish in my lounge room than spend any time on Mars with Elon Musk.) In the end, despite fantasies about moving elsewhere, I know I need to stay put, get to know my neighbours, and put all my energies, hope and trust in the community around me. (I guess I'm lucky, because I really like my neighbours.)

7 Take care of yourself

Put simply, thinking about climate change, talking about it with others and reading about it, not to mention experiencing it, involves an endless roller-coaster of emotion. Take care of yourself. Seek help if you need it. There are resources online and increasingly therapists specialising in eco-anxiety. Have a look at the resources section at the end of this book. And try to work out your own 'window of tolerance' (see pages 174–75).

8 Vote if you can

Most people in the world can't vote. Even in democracies like the United States, prisoners and ex-prisoners can't vote. In countries with elections but with electoral systems that are corrupt, where voting is difficult and even dangerous, the right to vote is an empty right. Given climate change is a global issue, given it will affect everyone in the world in some capacity, those of us who can vote in an open and legitimate democracy have a profound obligation to send a message through the ballot box that climate change requires immediate and effective action from government.

Green consumerism—buying environmentally friendly products and engaging in behaviour modification for the sake of the environment—is important, but until the politicians fear losing their jobs through not doing anything about climate change, we'll continue on this same, slow, inadequate path. Whatever political party you normally vote for, between elections send the message loud and clear to its leaders that climate change matters to you as a voter and a citizen. Protesting is one way of doing it, of course, or joining various campaigns by NGOs. But with all of the suggestions I've made in this chapter and previous ones, don't forget to tell your local politicians about what you're doing. If you're sending a message, make it personal and specific (standard form letters or emails are often ignored). It's not just ticking a box on a survey that says 'Yes, I'm concerned'. It's ticking a box on a ballot paper that says, 'You're out of a job if you don't act on this concern about climate change.'

9 Find your own climate story

At the beginning of this book, I wrote about the importance of stories of conviction and endurance as a way to connect people to climate change, making it personally relevant. Throughout the book I've brought you interviews with farmers, academics, artists, activists, museum curators, faith leaders and scientists as a way to understand how they use emotion in their work on climate and manage their own emotional reaction to climate.

Part of being able to talk about climate change is finding a way to tell your own climate story. How do you feel? How has that changed over time? What people, places and practices have made you engage with the issue? Sharing your own personal story about climate is one way of shaking up what has become a highly politicised and 'constipated' conversation. Hopefully it will elicit empathy and understanding from the person you're talking to as well, getting them to open up about their own story and why they feel the way they do about the climate.

But it's not just about 'your story', it's about amplifying the stories of other people in your community and around the world who are affected by climate change. It's the stories of Lavetanalagi Seru, of Chido, of the farmers I've met over the years, of the men and women of the Torres Strait, that I want to highlight in my work. Find your own climate storytellers and see what you can learn from them. American critic Rebecca Solnit writes that 'we are not very good at telling stories about a hundred people doing things'.[12] Which is a shame, she argues, because social change never comes from the actions of an isolated individual but from people connecting and acting with others in a coordinated push as a crowd united by the same values. We need to get better at telling hundreds of stories, stories about crowds in action.

As I finish this book, it's been a little over a year since I watched those climate kids protesting on the TV one Saturday morning and my world shifted on its axis. A year of writing, reading and reflecting on climate change has necessarily involved managing a perfect storm of emotions. Fundamentally, I've lost a strong confidence that the future will be better than the past and that my children will experience the same or even better opportunities than I did.

We often think, erroneously, of coming to terms with such a loss as going through the Elisabeth Kübler-Ross stages of grief—denial, anger, bargaining, depression and acceptance—neatly, one after the other (although, of course, Kübler-Ross herself acknowledged that we never experience these stages in a linear fashion). In truth, I think I've moved away from a kind of denial about climate change—that it was important but in a somewhat distant and remote way, not one that required me to sacrifice or reflect on my life in any meaningful fashion. But I now experience anger, depression, despair, guilt and fear, often and together, and in unexpected ways. I expect to for some time, with no obvious end in sight.

What I'm getting better at is returning to hope, the sceptical activism I talk about in this book, grounded in a firm belief in the power of groups of human beings to solve problems when determined enough to do so. As Greta Thunberg said in the midst of the COVID-19 pandemic, 'The coronavirus is a terrible event . . . But it also shows one thing: That once we are in a crisis, we can . . . act fast and change our habits and treat a crisis like a crisis.'[13]

And I keep returning to the words of writer Matthew Wilburn King: 'It's true that no other species has evolved to create such a large-scale problem. But no other species has evolved with such an extraordinary capacity to solve it, either.'[14]

And I keep going back, time and again, to love.

ALWAYS LOVE.

ACKNOWLEDGEMENTS

I am indebted to Per Espen Stoknes and George Marshall, whose work in the area of climate change communication has been so influential and guided me through the process of writing this book.

A special shout-out to Linh Do for connecting me with so many great climate communicators around the globe.

Thanks to Grant McDowell, Akufuna Muyunda, Ken Berlin, Stacie Paxton Cobos and Dan Kanninen for making time to talk and share their ideas with me.

Thanks to friends and colleagues who supported me through this process: Felicity Wade, Lyndon Schneiders, Damian Ogden, James Bradley, Owen Wareham and the great people at WWF, Blanche Verlie, Kelly Doust, Cecilia Anthony, Louise Wagner, Moksha Watts, Linda Scott, Deanne Head, Amy Stockwell, Sarah Macdonald, Carmen Lawrence and Eloise Spitzer. Thanks to the patience and understanding of my peers at work (Pino, Josh and Rea in particular). Thanks also to the Climate Reality teams, in both Australia and the United States. Thanks to the great people at the Yale Program, especially Lisa Fernandez, Anthony Leiserowitz, Parrish Bergquist, Seth Rosenthal and Matt Goldberg. Thanks also to Professor Carol Johnson and Professor Martha Augoustinos at Adelaide University.

Thanks to Isabella Bowdler for outstanding research assistance.

As always, my spectacular agent, Jeanne Ryckmans from Cameron's Management, cared about this project well beyond the confirmation of the publisher's contract.

The passion and dedication of the Murdoch Books team was second to none: Lou Johnson, Julie Mazur Tribe, Carol Warwick, Nicola Young and last but definitely not least Jane Morrow.

Thanks also to Daniel Yarrow and the extended Bowen and Yarrow clan for distracting the children while I wrote. And my mum and Graham for letting me move back home from time to time to write as well.

NOTES

Introduction: A change of heart

1. Rob Law, 'I have felt hopelessness over climate change. Here is how we move past the immense grief', *The Guardian*, 9 May 2019 <theguardian.com/commentisfree/2019/may/09/i-have-felt-hopelessness-over-climate-change-here-is-how-we-move-past-the-immense-grief>
2. Greta Thunberg, 'The disarming case to act right now on climate change', TEDxStockholm, November 2018 <ted.com/talks/greta_thunberg_the_disarming_case_to_act_right_now_on_climate_change>
3. Per Espen Stoknes, *What We Think About When We Try Not to Think About Global Warming: Toward a new psychology of climate action*, White River Junction, Vermont: Chelsea Green Publishing, 2015, p. 132
4. Stoknes, *What We Think About*, p. 133
5. Stoknes, *What We Think About*, p. 133
6. Thunberg, 'The disarming case to act'

Chapter 1: The problem with reason

1. Edward O. Wilson, *Half-Earth: Our planet's fight for life*, New York: W.W. Norton & Co., 2016, p. 1
2. Stoknes, *What We Think About*, p. 38–39; author's italics
3. Stoknes, *What We Think About*, p. 81
4. Stoknes, *What We Think About*, p. 81
5. Nathaniel Rich, *Losing Earth: The decade we could have stopped climate change*, New York: Picador, 2019, pp. 180, 200
6. Rich, *Losing Earth*, p. 5
7. George Marshall, *Don't Even Think About It: Why our brains are wired to ignore climate change*, London: Bloomsbury, 2014, p. 121

Chapter 2: Start being emotional

1. See 'Episode 42: Anthony Leiserowitz', CleanCapital <cleancapital.com/2019/03/episode-42-anthony-leiserowitz>
2. For more information, see 'Global warming's Six Americas', Yale Program on Climate Change Communication <climatecommunication.yale.edu/about/projects/global-warmings-six-americas>
3. Rich, *Losing Earth*, p. 5
4. Rich, *Losing Earth*, p. 42
5. Rich, *Losing Earth*, pp. 41–2
6. Rich, *Losing Earth*, p. 45
7. Mike Hulme, *Why We Disagree About Climate Change: Understanding controversy, inaction and opportunity*, Cambridge: Cambridge University Press, 2009, p. 364
8. Anthony A. Leiserowitz, 'American risk perceptions: is climate change dangerous?', *Risk Analysis*, 2005, vol. 25, no. 6, pp. 1433–42
9. Lorraine Whitmarsh, 'Scepticism and uncertainty about climate change: dimensions, determinants and change over time', *Global Environmental Change*, 2011, vol. 21, no. 2, pp. 690–700
10. Bruce Tranter, 'It's only natural: conservatives and climate change in Australia', *Environmental Sociology*, 2017, vol. 3, no. 3, pp. 274–85
11. Isobel Gladston & Trevelyan Wing, 'Social media and public polarization over climate change in the United States', Climate Institute (US), 27 August 2019 <climate.org/social-media-and-public-polarization-over-climate-change-in-the-united-states>
12. Bruce Stokes, Richard Wike & Jill Carle, 'Global concern about climate change, broad support for limiting emissions', Pew Research Center, 5 November 2015 <pewresearch.org/global/2015/11/05/global-concern-about-climate-change-broad-support-for-limiting-emissions>

13 See J. Marshall Shepherd, '3 kinds of bias that shape your worldview', TEDxUGA, March 2018 <ted.com/talks/j_marshall_shepherd_3_kinds_of_bias_that_shape_your_worldview>

Chapter 3: Green girls

1 See Danielle F. Lawson et al., 'Children can foster climate change concern among their parents', *Nature Climate Change*, 2019, vol. 9, pp. 458–62
2 Lawson et al., 'Children can foster climate change concern', p. 459
3 Lawson et al., 'Children can foster climate change concern', p. 460; my italics
4 Lawson et al., 'Children can foster climate change concern', p. 460
5 See Sifan Hu & Jin Chen, 'Place-based inter-generational communication on local climate improves adolescents' perceptions and willingness to mitigate climate change', *Climatic Change*, 2016, vol. 138, pp. 425–38
6 Hu & Chen, 'Place-based inter-generational communication', p. 428
7 Hu & Chen, 'Place-based inter-generational communication', p. 425
8 Hu & Chen, 'Place-based inter-generational communication', p. 436
9 Marshall, *Don't Even Think About It*, p. 187
10 Marshall, *Don't Even Think About It*, p. 191
11 See my 2019 Melbourne Sustainable Society Institute Oration, 'Renewing democracy in a time of environmental crisis' (at <sustainable.unimelb.edu.au/past-events/mssi-oration-with-rebecca-huntley>) for a discussion of this research
12 Marshall, *Don't Even Think About It*, p. 189
13 Marshall, *Don't Even Think About It*, p. 190
14 'Transcript: Greta Thunberg's speech at the U.N. Climate Action Summit', NPR, 23 September 2019 <npr.org/2019/09/23/763452863/transcript-greta-thunbergs-speech-at-the-u-n-climate-action-summit>

15 Katharine Wilkinson, 'How empowering women and girls can help stop global warming', TEDWomen 2018, November 2018 <ted.com/talks/katharine_wilkinson_how_empowering_women_and_girls_can_help_stop_global_warming>

16 Wilkinson, 'How empowering women and girls'.

17 Wilkinson, 'How empowering women and girls'.

18 Daniel A. Chapman, Brian Lickel & Ezra M. Markowitz, 'Reassessing emotion in climate change communication', *Nature Climate Change*, 2017, vol. 7, pp. 850–2

Chapter 4: Guilt

1 Daniel Szyncer, quoted in Eve Glicksman, 'Your brain on guilt and shame', 12 September 2019 <brainfacts.org/thinking-sensing-and-behaving/emotions-stress-and-anxiety/2019/your-brain-on-guilt-and-shame-091219>

2 Glicksman, 'Your brain on guilt and shame'. The author of this article refers to work done by researchers at the Ludwig Maximilian University in Munich

3 Stoknes, *What We Think About*, p. 60

4 Stoknes, *What We Think About*, p. 5

5 See Stokes et al., 'Global concern about climate change'

6 Marshall, *Don't Even Think About It*, p. 193

7 Jonas H. Rees, Sabine Klug & Sebastian Bamberg, 'Guilty conscience: motivating pro-environmental behaviour by inducing negative moral emotions', *Climatic Change*, 2015, vol. 130, pp. 439–52

8 Claudia R. Schneider et al., 'The influence of anticipated pride and guilt on pro-environmental decision making', 2017, *PLoS ONE*, vol. 12, no. 11, article no. e0188781 <doi.org/10.1371/journal.pone.0188781>

9 Hang Lu & Jonathon P. Schuldt, 'Compassion for climate change victims and support for mitigation policy', *Journal of Environmental Psychology*, 2016, vol. 45, pp. 192–200

10 Schneider et al., 'The influence of anticipated pride and guilt'

11 Ben Doherty & Michael Slezak, '"The island is being eaten": how climate change is threatening the Torres Strait', *The Guardian*, 13 July 2017 <theguardian.com/environment/2017/jul/13/the-island-is-being-eaten-how-climate-change-is-threatening-the-torres-strait>

12 Annah Piggott-McKellar, Karen Elizabeth McNamara & Patrick D. Nunn, 'Climate change forced these Fijian communities to move—and with 80 more at risk, here's what they learned', The Conversation, 30 April 2019 <theconversation.com/climate-change-forced-these-fijian-communities-to-move-and-with-80-more-at-risk-heres-what-they-learned-116178>

13 Lu & Schuldt, 'Compassion for climate change victims', p. 194

14 Lu & Schuldt, 'Compassion for climate change victims', p. 192

15 Lu & Schuldt, 'Compassion for climate change victims', p. 197

Chapter 5: Fear

1 David Wallace-Wells, *The Uninhabitable Earth: A story of the future*, New York: Penguin Press, 2019, p. 3

2 Per Espen Stoknes, 'How to transform apocalypse fatigue into action on global warming', TEDGlobal>NYC, September 2017, <ted.com/talks/per_espen_stoknes_how_to_transform_apocalypse_fatigue_into_action_on_global_warming?language=en>

3 Stoknes, *What We Think About*, p. 31

4 Hulme, *Why We Disagree About Climate Change*, p. 14

5 Hulme, *Why We Disagree About Climate Change*, p. 13

6 Hulme, *Why We Disagree About Climate Change*, p. 2

7 Stoknes, *What We Think About*, pp. 45–6

8 Stoknes, *What We Think About*, p. 40
9 Rich, *Losing Earth*, p. 40
10 Rich, *Losing Earth*, p. 112
11 Stoknes, *What We Think About*, p. 27
12 Stoknes, *What We Think About*, p. 44
13 See Allan Mazur, *True Warnings and False Alarms: Evaluating fears about the health risks of technology*, London: Routledge, 2004
14 Stoknes, *What We Think About*, p. 45
15 Stoknes, *What We Think About*, p. 44
16 Brigitte Nerlich & Rusi Jaspal, 'Images of extreme weather: symbolising human responses to climate change', *Science as Culture*, 2014, vol. 23, no. 2, pp. 253–76
17 Lisa Zaval et al., 'How warm days increase belief in global warming', *Nature Climate Change*, 2014, vol. 4, pp. 143–7
18 See Allan Mazur, 'Global warming in the fickle news', *Social Science and Public Policy*, 2019, vol. 56, pp. 613–19
19 Parrish Bergquist & Christopher Warshaw, 'Does global warming increase public concern about climate change?', 23 July 2018, <chriswarshaw.com/papers/ClimateOpinion_180322_public.pdf>
20 Bergquist & Warshaw, 'Does global warming increase public concern?', p. 2
21 Bergquist & Warshaw, 'Does global warming increase public concern?', p. 2
22 Rich, *Losing Earth*, p. 208
23 Saffron O'Neill & Sophie Nicholson-Cole, '"Fear won't do it": promoting positive engagement with climate change through visual and iconic representations', *Science Communication*, 2009, vol. 30, no. 3, pp. 355–79
24 Jochen Kleres & Åsa Wettergren, 'Fear, hope, anger, and guilt in climate activism', *Social Movement Studies*, 2017, vol. 16, no. 5, pp. 507–19; p. 507

25 Sarah Wolfe & Amit Tubi, 'Terror management theory and mortality awareness: a missing link in climate response studies?', *Climate Change*, 2019, vol. 10, article no. e566, p. 5

26 Christofer Skurka et al., 'Pathways of influence in emotional appeals: benefits and trade-offs of using fear or humor to promote climate change-related intentions and risk perceptions', *Journal of Communication*, 2018, vol. 68, no. 1, pp. 169–93

27 Maxwell Boykoff & Beth Osnes, 'A laughing matter? Confronting climate change through humor', *Political Geography*, 2019, vol. 68, pp. 154–63

28 See N. Biddle et al., 'Exposure and the impact on attitudes of the 2019–20 Australian bushfires', ANU Centre for Social Research and Methods, 2020 <csrm.cass.anu.edu.au/research/publications/exposure-and-impact-attitudes-2019-20-australian-bushfires-0>

29 Ben Doherty, 'Climate summit calls for urgent action after Australia's fire-hit summer', *The Guardian*, 15 February 2020 <theguardian.com/environment/2020/feb/15/climate-summit-calls-for-urgent-action-after-australias-fire-hit-summer>

Chapter 6: Anger

1 Ted Brader & Nicholas A. Valentino, 'Identities, interests, and emotions: symbolic versus material wellsprings for fear, anger, and enthusiasm', in W. Russell Neuman et al. (eds), *The Affect Effect: Dynamics of emotion in political thinking and behavior*, Chicago: University of Chicago Press, 2007, pp. 180–201, p. 183

2 Leonie Huddy, Stanley Feldman & Erin Cassese, 'On the distinct political effects of anxiety and anger', in Neuman et al. (eds), *The Affect Effect*, pp. 202–30, p. 206

3 Chapman et al., 'Reassessing emotion in climate change communication', p. 851

4   Will Coldwell, 'Anger is an energy: how to turn fury into a force for good', *The Guardian*, 13 May 2019, © Guardian Media Ltd 2020 <theguardian.com/lifeandstyle/2019/may/13/anger-interviews>
5   Kleres & Wettergren, 'Fear, hope, anger, and guilt in climate activism', p. 514
6   Marshall, *Don't Even Think About It*, p. 34
7   Marshall, *Don't Even Think About It*, p. 41
8   E.M. Cody et al., 'Climate change sentiment on Twitter: an unsolicited public opinion poll', *PLoS ONE*, 2015, vol. 10, no. 8, article no. e0136092 <doi.org/10.1371/journal.pone.0136092>
9   Ashley A. Anderson, 'Effects of social media use on climate change opinion, knowledge, and behavior', *Oxford Research Encyclopedia of Climate Science*, March 2017 <10.1093/acrefore/9780190228620.013.369>
10 Hang Lu & Jonathon P. Schuldt, 'Exploring the role of incidental emotions in support for climate change policy', *Climatic Change*, 2015, vol. 131, pp. 719–26, p. 724
11 Huddy et al., 'On the distinct political effects', p. 228
12 Stoknes, *What We Think About*, p. 179
13 Marshall, *Don't Even Think About It*, p. 42
14 Margaret V. du Bray et al., 'Emotion, coping, and climate change in island nations: implications for environmental justice', *Environmental Justice*, 2017, vol. 10, no. 4, pp. 102–107, p. 106
15 Stoknes, *What We Think About*, p. 179
16 Blanche Verlie, 'Bearing worlds: learning to live-with climate change', *Environmental Education Research*, 2019, vol. 25, no. 5, pp. 751–66
17 Jonathan Haidt, *The Righteous Mind: Why good people are divided by politics and religion*, New York: Penguin, 2012, p. 127
18 Haidt, *The Righteous Mind*, p. 127
19 Marshall, *Don't Even Think About It*, Chapter 9

Chapter 7: Denial

1. Stoknes, *What We Think About*, p. 16
2. Stoknes, *What We Think About*, p. 17
3. Stoknes, *What We Think About*, p. 10
4. Andrew J. Hoffman, 'Talking past each other? Cultural framing of skeptical and convinced logics in the climate change debate', Ross School of Business Working Paper no. 1154, February 2011 <hdl.handle.net/2027.42/83161>, p. 6
5. Stoknes, *What We Think About*, p. 82
6. Stoknes, *What We Think About*, p. 17
7. Stoknes, *What We Think About*, p. 74
8. Stephan Lewandowsky, John Cook & Elisabeth Lloyd, 'The "Alice in Wonderland" mechanics of the rejection of (climate) science: simulating coherence by conspiracism', *Synthese*, 2018, vol. 195, pp. 175–96
9. Aaron M. McCright & Riley E. Dunlap, 'Cool dudes: the denial of climate change among conservative white males in the United States', *Global Environmental Change*, 2011, vol. 21, no. 4, pp. 1163–72
10. Jonas Anshelm & Martin Hultman, 'A green fatwā? Climate change as a threat to the masculinity of industrial modernity', *International Journal of Masculinity Studies*, 2015, vol. 9, no. 2, pp. 84–96
11. Gordon Gauchat, 'Politicization of science in the public sphere: a study of public trust in the United States, 1974 to 2010', *American Sociological Review*, 2012, vol. 77, no. 2, pp. 167–87
12. Matthew Smith, 'International poll: most expect to feel impact of climate change, many think it will make us extinct', YouGov, 15 September 2019 <yougov.co.uk/topics/science/articles-reports/2019/09/15/international-poll-most-expect-feel-impact-climate>
13. Bruce Tranter, 'Climate scepticism in Australia and in international perspective', in S. Wilson & M. Hadler (eds), *Australian Social Attitudes IV*, Sydney: Sydney University Press, 2018, pp. 81–98

14 'Annual surveys of Australian attitudes to climate change', CSIRO, 20 June 2019 <csiro.au/en/Research/LWF/Areas/Pathways/Climate-change/Climate-attitudes-survey>

15 Pam Wright, '87 per cent of Americans unaware there's scientific consensus on climate change', Weather Channel, 11 July 2017 <weather.com/science/environment/news/americans-climate-change-scientific-consensus>

16 Hulme, *Why We Disagree About Climate Change*, p. 83

17 Stoknes, *What We Think About*, p. 14

18 Marshall, *Don't Even Think About It*, p. 36

19 Anna Rose, *Madlands: A journey to change the mind of a climate sceptic*, Melbourne: Melbourne University Press, 2012, p. 5

20 Rose, *Madlands*, p. 185

21 Rose, *Madlands*, p. 183

22 Rose, *Madlands*, p. 267

23 See, for example, Mark Hoofnagle, 'About—What is denialism?', Denialism (blog), 30 April 2007 <denialism.com/about>

24 Rose, *Madlands*, p. 127

25 Rose, *Madlands*, p. 143

26 Marshall, *Don't Even Think About It*, p. 4

Chapter 8: Despair

1 'The psychological effects of global warming', US National Wildlife Federation, 12 March 2012 <nwf.org/en/Educational-Resources/Reports/2012/03-12-2012-Psychological-Effects-Global-Warming>

2 Madeleine Thomas, 'Climate depression is for real. Just ask a scientist', Grist, 28 October 2014 <grist.org/climate-energy/climate-depression-is-for-real-just-ask-a-scientist>

3 Wallace-Wells, *The Uninhabitable Earth*, p. 207

4 Bruno Latour, *Down to Earth: Politics in the new climatic regime*, Cambridge: Polity Press, 2017, p. 6

5   Diana Younan et al., 'Long-term ambient temperature and externalizing behaviors in adolescents', *American Journal of Epidemiology*, 2018, vol. 187, no. 9, pp. 1931–41

6   For a collection of Nick Obradovich's scholarly works, see his Google Scholar search results at <scholar.google.com/citations?user=sAxggesAAAAJ&hl=en>

7   American Psychological Association Task Force on the Interface Between Psychology and Global Climate Change, 'Psychology and global climate change: addressing a multi-faceted phenomenon and set of challenges', American Psychological Association, 2009 <apa.org/science/about/publications/climate-change>. The American Psychiatric Association has released a similar report—see 'How extreme weather events affect mental health', American Psychiatric Association <psychiatry.org/patients-families/climate-change-and-mental-health-connections/affects-on-mental-health>

8   Latour, *Down to Earth*, p. 6

9   Thomas J. Doherty & Susan Clayton, 'The psychological impacts of global climate change', *American Psychologist*, 2011, vol. 66, no. 4, pp. 265–76

10 Doherty & Clayton, 'The psychological impacts of global climate change', p. 269

11 'David Buckel, prominent New York LGBT lawyer, dies after setting himself on fire', *The Guardian*, 15 April 2018 <theguardian.com/us-news/2018/apr/15/david-buckel-prominent-new-york-lgbt-lawyer-dies-after-setting-himself-on-fire>

12 Caitlin Fitzsimmons, '"It doesn't feel justifiable": the couples not having children because of climate change', *Sydney Morning Herald*, 22 September 2019 <smh.com.au/lifestyle/life-and-relationships/it-doesn-t-feel-justifiable-the-couples-not-having-children-because-of-climate-change-20190913-p52qxu.html?btis>

13 For more on this, see David Wallace-Wells' chapter 'Ethics at the end of the world' in *The Uninhabitable Earth*

14 H.E. Erskine et al., 'The global coverage of prevalence data for mental disorders in children and adolescents', *Epidemiology and Psychiatric Sciences*, 2017, vol. 26, no. 4, pp. 395–402

15 For Australia, see Mission Australia's Annual Youth Survey <missionaustralia.com.au/what-we-do/research-impact-policy-advocacy/youth-survey>. For the United Kingdom, see—from the UK arm of the YMCA—'A different world: the challenges facing young people', 2019 <ymca.org.uk/wp-content/uploads/2019/02/Challenges-facing-young-people.pdf>. For the United States, see the 2019 Pew study by Juliana Menasce Horowitz & Nikki Graf, 'Most U.S. teens see anxiety and depression as a major problem among their peers', Pew Center, 20 February 2019 <pewsocialtrends.org/2019/02/20/most-u-s-teens-see-anxiety-and-depression-as-a-major-problem-among-their-peers>

16 Erskine et al., 'The global coverage of prevalence data'

17 Susie E.L. Burke, Ann V. Sanson & Judith Van Hoorn, 'The psychological effects of climate change on children', *Current Psychiatry Reports*, 2018, vol. 20, article no. 35, p. 35

18 Blanche Verlie, 'The terror of climate change is transforming young people's identity', The Conversation, 15 March 2019 <theconversation.com/the-terror-of-climate-change-is-transforming-young-peoples-identity-113355>

19 Wallace-Wells, *The Uninhabitable Earth*, p. 11

20 See the section 'The Unsilent Majority' in my *Quarterly Essay* no. 73: *Australia Fair, Listening to the Nation*, 2019, Melbourne: Black Inc., p. 4

21 James Purtill, 'Breaking up over climate change: my deep dark journey into doomer Facebook', ABC Triple J Hack, 7 November 2019 <abc.net.au/triplej/programs/hack/breaking-up-over-climate-change-my-journey-into-doomer-facebook/11678736>

22 Daniel Kahneman, quoted in Marshall, *Don't Even Think About It*, p. 56

23 Marshall, *Don't Even Think About It*, p. 93
24 Terry Patten, 'Radical adaptation', *Dumbo Feather*, 2019, no. 61, p. 12
25 Raymond De Young, 'Transitioning to a new normal: how ecopsychology can help society prepare for the harder times ahead', *Ecopsychology*, 2013, vol. 5, no. 4, pp. 237–9, p. 239
26 Wallace-Wells, *The Uninhabitable Earth*, p. 8
27 Hannah Malcolm, 'Apocalypse soon: rejecting despair and denial about climate change', Theos, 9 October 2018 <theosthinktank.co.uk/comment/2018/10/09/apocalype-soon-rejecting-despair-and-denial-about-climate-change>
28 Jem Bendell, 'Deep Adaptation: a map for navigating climate tragedy', IFLAS Occasional Paper 2 <lifeworth.com/deepadaptation.pdf>
29 Bendell, 'Deep Adaptation'.
30 Bendell, 'Deep Adaptation'.
31 Kathryn Stevenson & Nils Peterson, 'Motivating action through fostering climate change hope and concern and avoiding despair among adolescents', *Sustainability*, 2016, vol. 8, no. 1, article no. 6

Chapter 9: Hope

1 'A global cooking community with an appetite for change', MAD, <madfeed.co/video/help-shape-the-future-of-mad-2>
2 Rich, *Losing Earth*, p. 77
3 Barbara L. Frederickson, 'Why choose hope?', *Psychology Today*, 23 March 2009 <psychologytoday.com/blog/positivity/200903/why-choose-hope>
4 Neil D. Weinstein, 'Unrealistic optimism about future life events', *Journal of Personality and Social Psychology*, 1980, vol. 39, no. 5, pp. 806–20
5 The Yale Program, as part of the Six Americas project, has found in numerous surveys that the majority of Americans think climate

change won't affect them in their lifetimes. In Australia at least, however, this is shifting over time. In a recent survey called 'Australia Talks', conducted by the national broadcaster, the ABC, more than 70 per cent of those surveyed believed that climate change was the most pressing issue impacting them personally (see Annabel Crabb, 'Australia Talks National Survey reveals what Australians are most worried about', ABC News, 8 October 2019 <abc.net.au/news/2019-10-08/annabel-crabb-australia-talks-what-australians-worry-about/11579644>). While the survey data on this remains mixed, the tendency for segments of the population to hold climate change at a distance, as not relevant to them in time and place, remains strong

6   Ezra M. Markowitz & Azim F. Shariff, 'Climate change and moral judgement', *Nature Climate Change*, 2012, vol. 2, pp. 243–7, p. 244
7   Marshall, *Don't Even Think About It*, pp. 148–9
8   Victoria Campbell-Arvai et al., 'The influence of learning about carbon dioxide removal (CDR) on support for mitigation policies', *Climatic Change*, 2017, vol. 143, pp. 321–36
9   See Stokes et al., 'Global concern about climate change'
10 P. Sol Hart & Lauren Feldman, 'The impact of climate change-related imagery and text on public opinion and behavior change', *Science Communication*, 2016, vol. 38, no. 4, pp. 415–41
11 Hang Lu, 'The effects of emotional appeals and gain versus loss framing in communicating sea star wasting disease', *Science Communication*, 2016, vol. 38, no. 2, pp. 143–69
12 Richard S. Lazarus, *Emotion and Adaptation*, New York: Oxford University Press, 1991, p. 285
13 Stoknes, *What We Think About*, p. 222
14 See 'Hope in the dark: untold histories, wild possibilities' (book information), Rebecca Solnit (author website) <rebeccasolnit.net/book/hope-in-the-dark-untold-histories-wild-possibilities>

15 Quoted in Laurie Mazur, 'Susanne Moser holds hope lightly', *Dumbo Feather*, 2019, no. 61, p. 85
16 Kleres & Wettergren, 'Fear, hope, anger, and guilt in climate activism', p. 513
17 Marshall, *Don't Even Think About It*, p. 27
18 Matthew H. Goldberg et al., 'Perceived social consensus can reduce ideological biases on climate change', *Environment and Behavior*, 2019 <doi.org/10.1177/0013916519853302>
19 Thanks to Carmen Lawrence for alerting me to this case study. See 'Flex your power', Environmental Policy Center <epolicycenter.org/flex-your-power>
20 Veronika Budovska, Antonio Torres Delgado & Torvald Øgaard, 'Pro-environmental behaviour of hotel guests: application of the theory of planned behaviour and social norms to towel reuse', *Tourism and Hospitality Research*, 2020, vol. 20, no. 1, pp. 105–16
21 Matt Apuzzo and David D. Kirkpatrick, 'Covid-19 changed how the world does science, together', *New York Times*, 1 April 2020 <nytimes.com/2020/04/01/world/europe/coronavirus-science-research-cooperation.html?referringSource=articleShare>
22 See Stokes et al., 'Global concern about climate change'

Chapter 10: Loss

1 Huntley, *Australia Fair*, p. 19
2 See Nick Baker, 'How a climate change study from 12 years ago warned of this horror bushfire season', SBS News, 6 January 2020 <sbs.com.au/news/how-a-climate-change-study-from-12-years-ago-warned-of-this-horror-bushfire-season>
3 Daniel Kahneman & Amos Tversky, 'Prospect theory: an analysis of decision under risk', *Econometrica*, 1979, vol. 47, no. 2, pp. 263–92
4 For more information about the test (and its drawbacks), see 'The Stanford marshmallow experiment: how self-control affects

your success in life', Effectiviology <effectiviology.com/stanford-marshmallow-experiment-self-control-willpower>

5   Marshall, *Don't Even Think About It*, p. 66
6   See James Fernyhough, 'Climate change on track to make world "uninsurable": IAG', *Australian Financial Review*, 15 November 2018 <afr.com/companies/financial-services/climate-change-on-track-to-make-world-uninsurable-iag-20181115-h17xu5>
7   Rose, *Madlands*, p. 296
8   See, for example, '6 global warming skeptics who changed their minds', The Week, 1 September 2010 <theweek.com/articles/491378/6-global-warming-skeptics-who-changed-minds>. See also Ross Pomeroy, 'Trump's NASA chief changed his mind on climate change. He is a scientific hero', Space.com, 12 June 2018 <space.com/40857-trumps-nasa-chief-changed-his-mind-on-climate-change-he-is-a-scientific-hero.html>
9   Stoknes, *What We Think About*, p. 111
10 Stoknes, *What We Think About*, p. 111
11 Susie Wang et al., 'Emotions predict policy support: why it matters how people feel about climate change', *Global Environmental Change*, 2018, vol. 50, pp. 25–40, p. 26
12 Glenn Albrecht, 'The age of solastalgia', The Conversation, 7 August 2012 <theconversation.com/the-age-of-solastalgia-8337>
13 Susanne C. Moser, 'Navigating the political and emotional terrain of adaptation: community engagement when climate change comes home', in S.C. Moser & M.T. Boykoff (eds), *Successful Adaptation to Climate Change: Linking science and policy in a rapidly changing world*, London: Routledge, 2013, pp. 289–305
14 'Dalai Lama says strong action on climate change is a human responsibility', *The Guardian*, 20 October 2015 <theguardian.com/environment/2015/oct/20/dalai-lama-says-strong-action-on-climate-change-is-a-human-responsibility>

15 'Pope Francis encyclical and climate change', Catholic Climate Covenant <catholicclimatecovenant.org/encyclical>

16 The Evangelical Environmental Network's website is creationcare.org

17 Matthew H. Goldberg et al., 'A social identity approach to engaging Christians in the issue of climate change', *Science Communication*, 2019, vol. 41, no. 4, pp. 442–63

18 Panu Pihkala, 'Eco-anxiety, tragedy and hope: psychological and spiritual dimensions of climate change', *Zygon*, 2018, vol. 53, no. 2, pp. 545–69

19 Marshall, *Don't Even Think About It*, p. 216

20 Andrew J. Hoffman, 'Climate science as culture war', Ross School of Business Working Paper no. 1361, June 2012 <hdl.handle.net/2027.42/136210>, p. 11

21 See 'Katharine Hayhoe', Nova, 3 April 2011 <pbs.org/wgbh/nova/article/katharine-hayhoe>

22 Jim Antal, *Climate Church, Climate World: How people of faith must work for change*, Lanham, Maryland: Rowman & Littlefield Publishers, 2018. Quoted in Tom Montgomery Fate, 'Do you believe in God? Then you have a moral duty to fight climate change, writes Jim Antal', *Chicago Tribune*, 18 April 2018 <chicagotribune.com/entertainment/books/ct-books-climate-jim-antal-0422-story.html>

23 Marshall, *Don't Even Think About It*, p. 225

24 Marshall, *Don't Even Think About It*, p. 225

25 See Austyn Gaffney, 'Ashes to ashes and into trees', *Sierra*, 5 January 2020 <sierraclub.org/sierra/ashes-ashes-and-trees>

26 Suzanne Kelly, *Greening Death: Reclaiming burial practices and restoring our tie to the earth*, Lanham, Maryland: Rowman & Littlefield Publishers, 2015

Chapter 11: Love

1. Stoknes, *What We Think About*, p. 90
2. Stoknes, *What We Think About*, p. 90
3. Wang et al., 'Emotions predict policy support', p. 26
4. Wang et al., 'Emotions predict policy support', p. 29
5. Wang et al., 'Emotions predict policy support', p. 25
6. Marshall, *Don't Even Think About It*, p. 3
7. See Jennifer Brown, 'What will make you believe in global warming? How about a life-altering flood, study asks', *Colorado Sun*, 20 June 2019 <coloradosun.com/2019/06/20/climate-change-beliefs-after-colorado-floods>. See also A.E. Albright & D. Crow, 'Beliefs about climate change in the aftermath of extreme flooding', *Climatic Change*, 2019, vol. 155, pp. 1–17
8. The Yale Program website has since changed. See the Johns Hopkins University website, where Smithson-Stanley is a PhD candidate, for this text: <snfagora.jhu.edu/person/lynsy-smithson-stanley>
9. See 'Two-thirds of North American birds are at increasing risk of extinction from global temperature rise', Audubon <audubon.org/climate/survivalbydegrees>
10. Report provided to author by Lynsy Smithson-Stanley
11. Guide provided to author by Lynsy Smithson-Stanley

Conclusion: Talk about climate change

1. Marshall, *Don't Even Think About It*, p. 82
2. Stoknes, *What We Think About*, p. 17
3. Goldberg et al., 'Perceived social consensus'
4. See Katharine Hayhoe, 'The most important thing you can do to fight climate change: talk about it', TEDWomen 2018, November 2018 <ted.com/talks/katharine_hayhoe_the_most_important_thing_you_can_do_to_fight_climate_change_talk_about_it>

5 Marshall, *Don't Even Think About It*, p. 237
6 Wallace-Wells, *The Uninhabitable Earth*, p. 227
7 See 'Tennis Australia commits to United Nations climate change action', Tennis Australia, 6 June 2019 <tennis.com.au/news/2019/06/06/tennis-australia-united-nations-climate-change>
8 Marshall, *Don't Even Think About It*, Chapter 42
9 Marshall, *Don't Even Think About It*, Chapter 42
10 Bendell, 'Deep Adaptation'
11 Laurie Mazur, 'Susanne Moser holds hope lightly', p. 86
12 Rebecca Solnit, 'A hero is a disaster', *Whose Story Is This?*, London: Granta, 2019
13 Laurie Goering, 'Greta Thunberg says coronavirus shows world can "act fast" on crises', Reuters, 25 March 2020 <reuters.com/article/us-health-coronavirus-climate-greta/greta-thunberg-says-coronavirus-shows-world-can-act-fast-on-crises-idUSKBN21B2K9>
14 Matthew Wilburn King, 'How brain biases prevent climate action', BBC Future, 8 March 2019 <bbc.com/future/article/20190304-human-evolution-means-we-can-tackle-climate-change>

SUGGESTED READING AND RESOURCES

Here is a list of resources, writers and thinkers I have found useful, as well as the details of all the people I interviewed for this book.

People

Jim Antal: @JimAntal, jimantal.com

David Finnigan: davidfinig.com

Chido Govera: @chygovera, thefutureofhope.org

Tanya Ha: @Ha_Tanya

Mary Annaïse Heglar: @MaryHeglar

Daisy Jeffrey: @DaisyJeffrey2. See also schoolstrike4climate.com

Anthony Leiserowitz: @ecotone2. Visit the Yale Program at climatecommunication.yale.edu, and check out the Climate Connections podcast at yaleclimateconnections.org

George Marshall: @climategeorge, climateconviction.org. Read *Don't Even Think About It: Why our brains are wired to ignore climate change*

Miranda Massie: @MirandaKSMassie. Visit the Climate Museum at climatemuseum.org

Anna Oposa: @annaoposa and annaoposa.ph

Cassia Read: @cassiaread, climateflags.org

Anna Rose: annarose.net.au

Lavetanalagi Seru: @lagiseru, facebook.com/Alliance4FutureGenerations

J. Marshall Shepherd: @DrShepherd2013

Per Espen Stoknes: @estoknes, stoknes.no. Also read his book *What We Think About When We Try Not to Think About Global Warming: Toward a new psychology of climate action*

Katharine Wilkinson: @DrKWilkinson, kkwilkinson.com

Books

Active Hope: How to face the mess we're in without going crazy, by Joanna Macy & Chris Johnstone

Climate: A new story, by Charles Eisenstein

The Climate Change Playbook: 22 systems thinking games for more effective communication about climate change, by Dennis Meadows, Linda Booth Sweeney & Gillian Martin Mehers

Ecoliterate: How educators are cultivating emotional, social, and ecological intelligence, by Daniel Goleman, Lisa Bennett & Zenobia Barlow

The Parents' Guide to Climate Revolution: 100 ways to build a fossil-free future, raise empowered kids, and still get a good night's sleep, by Mary DeMocker

The Righteous Mind: Why good people are divided by politics and religion, by Jonathan Haidt

Thinking, Fast and Slow, by Daniel Kahneman

Websites

The Climate Reality Project
climaterealityproject.org

Deep Adaptation Forum
deepadaptation.info

Good Grief Network
goodgriefnetwork.org

MAD
madfeed.co

Peak bodies of mental health professionals in various countries have released helpful guides on climate change and mental health, including:

American Psychological Association
apa.org/news/press/releases/2017/03/mental-health-climate.pdf

British Psychological Society
bps.org.uk/topics/nature-and-environment

Australia Psychological Society
psychology.org.au/for-the-public/Psychology-Topics/Climate-change-psychology

Podcasts

Climate Conversations
This weekly MIT podcast focuses mostly on research and activism around climate solutions, but guests often discuss climate change communication.

Hello Climate Change
Amy Kalisher describes herself as an ordinary white middle-class American waking up to climate change. This podcast looks at her and her friends and family, and their efforts to reclaim their power as citizens through climate action.

Hot Take
Mary Annaïse Heglar and Amy Westervelt look at the way climate change is talked about in the media and the movement, and take a vitally important intersectional approach, highlighting the perspectives and activism of people of colour.

Mothers of Invention
Former Irish president Mary Robinson and comedian Maeve Higgins created this uplifting podcast featuring women from around the world working for climate justice.

My Climate Journey
A self-described recovering software entrepreneur goes on a quest to reorient his career around helping to solve climate change.

Movies

2040
Damon Gameau's vivid, positive and sometimes funny look at the future if we embraced the already existing and best solutions for climate change.

The Anthropologist
A film that follows the adventures of environmental anthropologist Susan Crate and her somewhat reluctant teen daughter Katie as they visit Indigenous communities around the world threatened by climate change.

The Biggest Little Farm
A documentary maker and a chef start a farm on 80 hectares (200 acres) outside Los Angeles. Their heroic attempt to farm in tune with the natural ecosystem is heartwarming and heartbreaking in equal measure.

Carbon Nation
An optimistic film looking at the solutions available to us to deal with climate change. It tries to do so in a non-preachy, non-partisan way, showing how acting on climate change can also boost the economy, and improve health and national security.

The Island President
A film about President Mohamed Nasheed of the Maldives, the lowest lying country in the world, as he tries to keep his country from being swallowed up by the rising tide.

The Magnitude of All Things
Jennifer Abbott looks at the emotional and psychological aspects of the climate crisis. Travelling around the world, she explores the relationship between grief and hope.

INDEX

A
aggressive behaviours 153
'Alarmed' population segment 33
Albrecht, Glenn 204
Alliance for Future Generations 84
American Psychological Association 154
Anderson, Ashley 122–3
anger
   behaviours linked to 118–20
   definition 118
   link between anger and fear 118
   link between anger and guilt 119
   as motivating force for climate activists 120–1, 129
   negative aspects of 124–7
   positive and constructive aspects of 119–20, 121
   role and effect of in talking about climate change 126–7
Anshelm, Jonas 138
Antal, Jim 208–15

anti-feminism and climate change denial, links between 138
anxiety epidemic in young people 156–8
see also mental health issues
apocalyptic scenarios
   apocalypse fatigue 93, 113
   De Young, Raymond, on 164
   Leiserowitz, Tony, on 32–3
   Rich, Nathaniel, on 35
   Stoknes, Per Espen, on 10–11
   Wallace-Wells, David, on 91, 92–4
   young men's perceptions of 98
   see also Deep Adaptation; doomers and doomer groups
Ardern, Jacinda 50
Aspen Global Change Institute 30–1
Attenborough, David 19
attitudes to climate change, segmentation of 33–4
Australian Youth Climate Coalition 143

B
backlash against climate change messages and messengers 17, 47
Bangladesh 99, 126
behaviour change by individuals *see* individual action to reduce climate change
Bendell, Jem 165–7, 240
benefits, perceived, of climate change 192
Bergquist, Parrish 101
biases, cognitive 42–3
Bird, John 120
birds in case study of climate change communication 223–30
BirthStriker movement 156
blame-shifting 74–5
Booker, Christopher 146
Bottura, Massimo 177
Branson, Richard 186
Bray, Margaret V. du 127
Buckel, David 155
burials, environmentally friendly 215–16
Bush, George W. 201
bushfires in Australia, 2019–20 summer
   capacity of to become teachable moments 202
   as experience of loss 195–7
   as parallel to book *The Uninhabitable Earth* 92
   as possible tipping point on climate change attitudes 100–1, 111–12, 125
   as trigger for anger and blame 125
business agenda 199

C
California 203
Campbell-Arvai, Victoria 186–7
Cato Institute 135
'Cautious' population segment 34
CERN (European Organization for Nuclear Research) 146
Chang, David 177
Chapman, Daniel A. 65, 119
Chen, Jin 55–6
children
   as climate change activists *see* student protests
   role of in public persuasion campaigns 58–62
Clayton, Susan 154
climate change, causes and effects of 8, 17, 20
   see also scientific method

climate change denial and deniers
  active compared with passive denial 132–3
  arguing with deniers, pros and cons of 142–3, 148
  comforting vision presented by deniers 131–2
  definitions 132–3
  demographics of deniers 138–9
  denial as marker of Republican allegiance 38, 200
  denial as self-protection mechanism 136–8, 200
  deniers' antagonism towards environmentalists 74, 146
  deniers compared with doomers 163
  fear as driver of denial 135–6
  I Can Change Your Mind About Climate Change (TV documentary) 144–8
  prevalence of in populations 139–4
  professional compared with amateur deniers 134
  sceptics compared with deniers 132
  silence as symptom and cause of 233–4
  see also climate change scepticism and sceptics
climate change scepticism and sceptics 133–5
  change of mind by sceptics 201
  I Can Change Your Mind About Climate Change (TV documentary) 144–8
  overcoming scepticism through connection with local concerns 54
  prevalence of in developed countries 40–1
  scepticism as marker of conservative allegiance 38–9, 138, 200–1
  scepticism from religious denominations 207
  sceptics compared with deniers 132
  see also climate change denial and deniers
climate change as social, cultural and political phenomenon 36–43, 49, 121–4
  see also political dimensions of climate change
'climate change' terminology, compared with 'global warming' 101, 201
'climate depression' 154
  see also mental health issues

Climate Flags project 168–73
Climate Institute (US) 39
Climate Museum (New York) 23–7
Climate Reality conference and training course (2019) 2, 20
   Gela, Fred 81–2
   Oposa, Anna 115
   Webb, Torres 82
climate scientists
   emotional responses experienced by 219–20
   mental health issues experienced by 151–2; *see also* mental health issues
Codling, Joanne 145–6
Cody, Emily 122
coffee cups, single-use versus reusable 68–9, 72
cognitive biases 42–3
cognitive dissonance 42
'collapse porn' 93
collective action 189–90
   see also personal action to reduce climate change
collective responsibility 71, 77
   see also blame-shifting; personal action to reduce climate change
communication on climate change
   communication compared with further research, relative importance of 20
   framing around objects of care *see* objects of care
   messenger compared with message, relative importance of 32
   National Audubon Society case study 223–34
   negative language, prevalence of 201–2
   positive framing, importance of 218
   reason versus emotion and worldview as basis of 13–14, 21–2, 28, 64; *see also* climate change denial and deniers
   role of religious leaders and messages 208–11, 227
   see also science communication; talking about climate change in daily life
'community', significance of 223, 241–3
compassion, role and effect of in talking about climate change 86–8

'Concerned' population segment 34
confirmation bias 42
conservative voters, communicating with
  denial and scepticism as markers of conservative allegiance 38, 200; *see also* climate denial and deniers
  National Audubon Society case study 225–30
control, lack of, as agent of despair 158–9
Cox, Brian 19
cultural dimensions of climate change *see* climate change as social, cultural and political phenomenon
Cyprus 127

D
Dalai Lama 206
death, effect of feelings about, in shaping response to climate change 103–4
Deep Adaptation 165–7
deficit model of science communication *see* science communication

denial *see* climate change denial and deniers
despair 158–61, 173–5
  constructive aspects of 165–8
  definition 158
  see also emotional resilience, building; mental health issues
De Young, Raymond 164
DiCaprio, Leonardo 74
'Disengaged' population segment 34
'Dismissive' population segment 34
divestment from fossil fuels 209
Doherty, Thomas J. 154
doomers and doomer groups 156, 160–1, 163, 165
'Doubtful' population segment 34
drought 87, 112, 143, 153, 180–1
du Bray, Margaret V. 127
Dunning-Kruger effect 42

E
eco-anxiety 154–5
  see also mental health issues
eco-burials 215–16
eco-guilt 69–70

education, formal, on climate change
  in school 22
  as pathway to intergenerational communication 52–5, 55–6
emotion versus reason as basis of persuasion 13–14, 19, 22–3, 28
emotional resilience, building 168, 174–5, 243
  Climate Flags project 168–73
  in face of climate inaction 117–18, 127–8, 128–9
  lessons from religious faiths 205–15
  specific strategies for 176
  window of tolerance 174–5
  see also mental health issues; hope
emotional responses to climate change see specific emotions
environmental grief 154, 204
  see also mental health issues; loss
environmentalists and environmental movement
  antagonism towards 74
  conflation of with climate change activism 73–4
  negative perceptions of 73–4

European Organization for Nuclear Research (CERN) 146
Evangelical Environmental Network 206
Evans, David 146
evolutionary psychology
  care for children, role of 60
  collective action, role of 190
  fear, role of 96–8
  guilt, role of 70
  human response to climate change 16, 95
  shame, role of 70
Extinction Rebellion 121
  compared with doomer groups 160–1
  links with Deep Adaptation 167
ExxonMobil 89

F
family-based discussions about climate change 52–7
farming 181, 182
'Fatalists' population segment 35
fear
  fear of death 103–4
  fear as driver of climate change denial 135–6

link between fear and anger
118, 118–19
role and effect of in talking
about climate change
94–105, 106–10, 113
Feldman, Lauren 187–8
Fiji 83–6, 126, 127, 139
Finnigan, David 108–10
Flannery, Tim 106–7, 173
food security 177–8
Franzen, Jonathan 162–3
Future of Hope 179
future scenarios, building a
positive picture of 32–3
see also apocalyptic scenarios;
Deep Adaptation

G
Garnaut, Ross 197
Gela, Fred 81–3
gender issues
gender equity as pathway to
reducing global emissions
62–4
gender gap in views about
climate change 46
girls
effectiveness of teen girls as
climate change campaigners
52–5, 64
role of girls in public persuasion
campaigns 58–60
Gladston, Isobel 39
'global warming' terminology,
compared with 'climate
change' 101, 201
Goldsmith, Zac 145
Gore, Al
as Climate Reality founder 20
criticism of behaviour of 74
Inconvenient Truth, An, impact
of 38
as young senator 30
Govera, Chido 177–83
Great Barrier Reef 4–5, 20, 203,
220
Guariglia, Justin Brice 26
guilt
author's feelings of 80–1, 88
climate change denial as
response to 136–7
compared with shame 70
definition 69
link between guilt and anger
119
links to prosocial behaviour 71
purpose 69–70
role and effect of in talking
about climate change 70–4,
77–80, 85–6, 88–9

H

Haidt, Jonathan 128–9
Hansen, James 30
Hart, P. Sol 187–8
Hayhoe, Katharine 208, 234–5
Heartland Institute 135
hoax, perception of climate change as 38
   see also climate change denial and deniers
Hoffman, Andrew (Andy) 135, 208
home, human connections with 205
Hoofnagle, Chris 147
Hoofnagle, Mark 147
hope 183–93
   absence of correlation with despair 173
   as response to tragedy typical of religious leaders 213–14
   role and effect of in talking about climate change 103, 241
'household hints' for addressing climate change 76–7
Huddy, Leonie 125
Hulme, Mike 37, 94, 142
Hultman, Martin 138
human characteristics, understanding of as key to climate change action 15, 31–2
humour, role and effect of in talking about climate change 118, 106–10
Hurricane Sandy 23
Hu, Sifan 55–6

I

images of climate change 99–100, 103, 187–8
Inconvenient Truth, An (documentary film) 38
Indigenous communities, resilience of 162
individual action to reduce climate change
   author's 2–3
   catalyst for 3, 23–4
   compared with policy framework 76–7
information-deficit model of science communication *see* science communication
insurance industry 200
intergenerational influence 48, 52–4, 55–7

intergenerational responsibility 1–2, 4–5, 58–61, 104, 105
Intergovernmental Panel on Climate Change (IPCC) 4, 99

J
Jaspal, Rusi 99
Jeffrey, Daisy 45–51

K
Kahneman, Daniel 161, 197
Kelly, Suzanne 216
King, Matthew Wilburn 247
Kleres, Jochen 103, 120–1, 189–90
Kyoto Protocol 38

L
Latour, Bruno 91, 153, 154
Law, Rob 5
Lawrence, Carmen 219
Lawson, Danielle 52–3
Leiserowitz, Tony 29–33, 37, 38, 105, 185
Lewandowsky, Stephan 137
Limbaugh, Rush 134
Lindzen, Richard 146
listening, importance of 237
Lomborg, Bjørn 146

loss
  author's experience of 203, 246
  definition 197
  framing of climate change discussion in terms of 201–2
  of home 204–5
  'marshmallow experiment' 198
  prospect theory 197–9
  solastalgia 204
love
  as influence on climate change beliefs and behaviour 217–18
  in terms of 'objects of care' 219–30
Lu, Hang 80, 86–8, 124, 188
Luntz, Frank 201

M
MAD Sydney 177–8
Madrid climate change conference (2019) 46
Malcolm, Hannah 164
Maldives 203
Marshall, George
  on blame 126, 130
  on deniers 143
  on human instinct to defend tribe 222–3

on lessons from faith
　　　　communities 214
　　on optimism 186
　　on personal-responsibility focus
　　　　of campaigns 76
　　on prospect theory 198
　　on religions 207
　　on role of young girls in
　　　　behaviour change
　　　　campaigns 58–60
　　on social conformity 190
　　on social media 122
　　on solutions 161–2
　　on talking effectively about
　　　　climate change 236
　　on worry 148
'marshmallow experiment' 198
mass extinction, sixth 8
Massie, Miranda 22–8, 162
media reporting of climate change
　　49, 99–100, 187–8
mental health issues 152–8, 168,
　　243
　　climate scientists' experiences of
　　　　151–2
　　pre-traumatic stress disorder
　　　　151–2
　　as rational response to climate
　　　　change 155

　　see also emotional resilience,
　　　　building; grief; loss
messenger and message, relative
　　importance of 57
Minchin, Nick 144–8
Monckton, Christopher 143
Morano, Marc 143, 146
Morrison, Scott 207
Moser, Susanne 189, 204–5
Musk, Elon 186

N

National Audubon Society
　　225–30
negative language 201–2
Nerlich, Brigitte 99
New Zealand 127
Nicholson-Cole, Sophie 103
Nye, Bill 19

O

'objects of care', love as
　　understood in terms of
　　219–30, 239–40
Obradovich, Nick 153–4
O'Neill, Saffron 103
Oposa, Anna 40–1, 115–18
optimism bias 184–6, 198
　　see also prospect theory
ozone depletion 95–6

P

Pacific Islander nations, effects of climate change on 83
  see also Fiji; Torres Strait Islands
Paris Agreement 211
Parks Victoria 170
Parmesan, Camille 151
Patten, Terry 163–4
Pauli, Gunter 178
Philippines 115–18, 139
Pihkala, Panu 207
plastic waste 115
polar bear as climate change symbol 99, 117
political dimensions of climate change
  correlation between political allegiance and climate change views 38–40, 200
  compassion, role of in cutting through partisan divisions 87
  intergenerational conversations, role of in cutting through partisan divisions 52–7
  see also climate change as social, cultural and political phenomenon
pre-traumatic stress disorder 151–2
pride, effect of on environmental decision-making 79, 80
Project Drawdown 63
prospect theory 197–9
  see also optimism bias
protests by school students see student protests
psychology of climate change perceptions 41–3
  see also specific emotions
Purtill, James 160

Q

Quadrant magazine 146

R

Read, Cassia 168–73
reason versus emotion as basis of persuasion 13–14, 19, 22–3, 28
Redzepi, René 177
religious beliefs, traditions and institutions 205–15
  descriptions of the end of the world in scriptures 164
  religion-based statements on need for environmental action 116, 206, 206–7, 227, 230

role of religious leaders in environmental action 208–13
as solace in grief 205
responsibility for action on climate change
  collective action 189–90
  individual consumers as agents of change 76–7; *see also* personal action to reduce climate change
responsibility for causing climate change
  climate change denial response 136–8
  collective responsibility 71, 77; *see also* blame-shifting
  responsibility as precursor to guilt 72–3
  wealthy countries compared with developing countries 72, 75–6, 83, 126–7
Rich, Nathaniel 17, 34–6, 102, 183
Rio de Janeiro 203
risk vividness 97–8
Rose, Anna 143–8, 193, 201

S
Save Philippine Seas 116
scapegoating 80
scepticism *see* climate change scepticism and sceptics; climate change denial and deniers
School Strike 4 Climate Australia 46
  see also student protests
Schuldt, Jonathon P. 80, 86–8, 124
science communication
  backlash against messengers 17
  failure of in relation to climate change 16–17, 17, 19
  human characteristics affecting message reception 15
  information-deficit model 14, 15–16, 16–17, 21–2
  persuasive capacity of famous communicators 19
  reason versus emotion and worldview as basis of 13–14, 21–2, 28; *see also* climate change denial and deniers
  role of climate scientists 19, 21
  role of politicians 19
  social approach 22; *see also* Climate Museum (New York)

storytelling and personal
information, absence of,
from 18–19
Yale Program on Climate
Change Communication
29–33
see also communication on
climate change; talking
about climate change in
daily life
scientific method 133–4
definition 17–18
uncertainty as characteristic of
142
Seru, Lavetanalagi 83–6, 127, 204
shame
author's feelings of 88
compared with guilt 70
links with antisocial behaviour
71
purpose 69–70
role and effect of in talking
about climate change 70–4,
74–5, 75–6, 77–80, 88–9
Shepherd, J. Marshall 41–2, 43
Siegel, Dan 174
Six Americas study 33–4
Smithson-Stanley, Lynsy 224–30
social dimensions of climate
change *see* climate change as

social, cultural and political
phenomenon
social media
role of in polarisation on
climate change 39, 121–3
temperature–sentiment
correlation 153–4
solastalgia 204
Solnit, Rebecca 188–9, 245
Stoknes, Per Espen
on anger 127
on blame 126
on denial and deniers 132, 136,
143
on information-deficit model
of science communication
15, 16
on intangibility of climate
change threat 95
on loss 201–2
on positive framing in
communication 218
on provoking 'shame' response
to communication 74, 75
on risk 97
on scepticism 133, 188
on storytelling 10–11
storytelling as communication
tool 10–11, 27–8, 244–5

case study of Climate Museum (New York) 23–7
communication failures caused by absence of 18–19
student protests
  academic research on 157–8
  author's response to 1–2, 4, 5
  Jeffrey, Daisy, as organiser of 45–51
  Law, Rob, response to 6
suicide
  association of with despair 158
  by environmentalist David Buckel 155
  see also mental health issues
Suzuki, David 19
symbols of climate change 99–100
Sznycer, David 69–70

T
talking about climate change in daily life 231–4
  how-to summary 235–6
  nine tips 237–45
  starting points 235–7
  see also communication on climate change; science communication

technological solutions to climate change 186–7
temperature rises
  as driver of behaviour changes 153–4
  as driver of climate change concern 101–2
Tennis Australia 238
theatre as climate change communication medium 108–10
Thompson, Emma 74
Thunberg, Greta 57–8, 61–2, 158, 246
  as figurehead of teenage girl climate activists 52
  on hope 10
  at Madrid climate talks 46
  TED talk 10, 12
'tipping point', Australian 2019–20 bushfires as 100, 111–12
Titley, David 145
Torres Strait Islands 81–4, 126, 203
toxic knowledge 152
Tranter, Bruce 140
Trump, Donald 57, 211
Tubi, Amit 104–5
Tversky, Amos 197

Twitter wars on climate change 121–2
  see also social media
Tyson, Neil deGrasse 19

U
union movement, Australian 199–200
United Church of Christ 208

V
Venice 203
Verlie, Blanche 128, 157–8
voting as climate change activism 243–4

W
Wallace-Wells, David 91, 152, 159, 164–5, 237
Wang, Susie 203
Warshaw, Christopher 101
weather events 100–2, 112, 180
  images of 99–100
  perception of as divine 94
  see also temperature rises
Webb, Torres 82
Weinstein, Neil 184
Wettergren, Åsa 103, 120–1, 189–90
white privilege 162, 163

Wiesel, Elie 219
wildfires see bushfires in Australia, 2019–20 summer
Wilkinson, Katharine 63
Wilson, Edward O. 15, 91
window of tolerance 174–5
Wing, Trevelyan 39
'woke', characterisation of environmental messages as 74
Wolfe, Sarah 104–5
women and girls, empowerment of, to reduce global emissions 62–4
  see also gender issues; girls
World Climate Conference (1979) 36
worry, role and effect of in talking about climate change 105–6
WWF Australia 60, 111–12

Y
Yale Program on Climate Change Communication 29–30, 224

Z
Zero Emissions Research and Initiatives Foundation 178
Zimbabwe 178–81